Insects of Texas: A PRACTICAL GUIDE

Number Thirty-nine:
W. L. Moody Jr. Natural History Series

ATM nature guides

Insects of Texas

A PRACTICAL GUIDE

David H. Kattes

TEXAS A&M UNIVERSITY PRESS COLLEGE STATION

Manufactured in China by
Everbest Printing Co., through
Four Colour Print Group

First edition

This paper meets the requirements
of ANSI/NISO Z39.48-1992
(Permanence of Paper).
Binding materials have been
chosen for durability.

LIBRARY OF CONGRESS CATALOGING-IN-PUBLICATION DATA

Kattes, David Hugh, 1952–
Insects of Texas : a practical guide / David H. Kattes.
p. cm. — (W. L. Moody Jr., natural history series)
Includes index.
ISBN–13: 978-1-60344-082-0 (flexbound with flaps : alk. paper)
ISBN–10: 1-60344-082-8 (flexbound with flaps : alk. paper)
1. Insects—Texas—Identification.
2. Arthropods—Texas—Identification.
I. Title. II. Series.
QL475.T4.K38 2009
595.709764—dc22 2008023996

Contents

Preface

"What kind of bug is this?" As an entomologist and teacher, I often hear this question. Typically, I respond with a common name like "mud dauber wasp" and, if the questioner seems truly interested, a somewhat more detailed description of the specimen's physical characteristics and biology. Occasionally, I ask myself, "How do I really know what insect that is?"

Insect taxonomy is a very tedious, meticulous science, and most entomologists spend a good portion of their career studying only a very small number of insects. When I studied introductory insect taxonomy at Texas A&M University in 1973, my professor, Horace Burke, required us to do a lot of memorizing. We memorized antennal types, tarsal formulas, number of ocelli, wing venation, and a myriad of other physical, biological, and ecological traits that distinguish a particular group. I have since forgotten many of the details, but I can still identify the more common groups by using readily distinguishable field characteristics.

Although the number and diversity of insect species can be overwhelming, generalizations can often be made about groups of related insects. This book was written for people who have little or no formal training in insect taxonomy but who need or desire to understand more about the classification of insects. This is not an authoritative or definitive book on insect taxonomy, but rather a guide to the more recognizable identifying characteristics of many common insect orders and families in Texas. For example, although Ross Arnett listed 113 families of beetles in his book *A Handbook of the Insects of America North of Mexico* (Gainesville, Fla.: Sandhill Crane Press, 1993), he suggested that 70 percent of all beetle species occur in only 12 families.

Because this guide describes groups of related insects based on their common and most easily observed field characteristics, I hope readers will find it helpful in sorting through the countless number of insects they encounter. Once the specimen is identified as belonging to a certain group, the reader may consult other references for more detailed information.

I would like to thank Ann Pawlak and Evelyn Kattes for reviewing early drafts of the manuscript and offering very constructive criticism. I am also indebted to Dr. Jeanelle Barrett and Shayna Dunn for developing the pronunciation guide and to Shayna for her helpful review of later drafts. Lastly, I want to dedicate this work to my wife, Molly, for all her patience, encouragement, and love.

Introduction

As humans, we tend to categorize objects and ideas into logical groups. Classification schemes can be based on use, size, color, shape, or a myriad of other methods. In my workshop, I have an area where I store various fasteners. In one bin, I have nails; in another, screws; and in still another, bolts. Even within these groups, I have additional divisions. I have separate bins for different types and sizes of bolts, such as $\frac{1}{8}$-inch hex bolts, $\frac{1}{4}$-inch carriage bolts, and so forth. Grouping similar objects enables us to find a particular item of interest quickly and efficiently.

Entomologists disagree about the number of insect species currently living on earth; nevertheless, with estimates ranging from 2 million to 30 million species, it is fair to say we personally can be familiar with only a very, very small number of insects. However, in many cases we can recognize groups of insects with morphological, biological, and ecological similarities.

Systematics, the study of the diversity of living organisms and the relationships between them, is a cornerstone of biology. Taxonomy, the science of describing, classifying, and naming organisms, is a subdiscipline of systematics that attempts to group living things into a logical scheme. Taxonomy is not a new concept. In Genesis 2:19–20, the first book of the Bible, God allowed Adam to name all the living creatures on earth. In the fourth century B.C., Aristotle developed the first known scientific classification system.

Carl Linnaeus, an eighteenth-century Swedish botanist, laid the foundation for the taxonomic system we now use. This system groups organisms according to their similarities from broadly similar to very similar. A simplified hierarchical system (from broad to specific) commonly used today is similar to the following:

Kingdom (broadest grouping)
  Phylum
    Class
      Order
        Family
          Genus
            Species (most specific grouping)

The fundamental unit of this system is the species, which may be defined as a group of organisms that are very similar in appearance, physiology, genetics, and ecology. Each species is assigned to a broader group called a genus that is composed of similar species. The members of a genus are grouped with other similar genera to form a broadly related family and so forth. To help you recognize group levels, this book uses color to distinguish families, orders, and classes. The above depiction of the hierarchical classification system is simplified. There are typically subdivisions at each rank; for example, in some cases, there may be superfamily, family, and subfamily groups.

Taxonomy is not static. As scientists learn more about an organism or group of organisms, the taxonomy is changed to reflect this new knowledge. A species may be reclassified in a different genus. What were originally believed to be two or more species may actually be the same, with the apparent differences due to geographic or other forms of isolation.

Data interpretation by various scientists may also influence the classification system. Some taxonomists, known as "splitters," separate groups based on several observed differences. Taxonomic "lumpers," however, may view these same differences as less important than some overriding major characteristic and thus have fewer groups.

What Is an Insect?

Insects are members of the kingdom Animal and phylum Arthropoda. Arthropods have a hardened shell called an exoskeleton, paired jointed appendages that include legs, fingerlike projections near the mouth called pedipalps, antennae, and other structures. These typically small animals are bilaterally symmetrical; if you were to draw a line down the middle of the organism, appendages on both sides of the line would be identical.

The phylum Arthropoda is divided into five classes: (1) Arachnida (spiders, scorpions, and others); (2) Crustacea (sowbugs, crayfish, and others); (3) Diplopoda (millipedes); (4) Chilopoda (centipedes); and (5) Hexapoda (insects). The hexapods are generally easily separated from the other classes by having three body segments, six legs, and one pair of antennae; they are the only class that may possess wings.

Internally, insects have a tubular gut with a mouth and an anus. Their nervous system consists of a brain located in the upper portion of the head. The brain is connected to a nerve cord located on the ventral, or underside, of the organism below the gut.

Insects have an open circulatory system. Rather than being pumped through veins and arteries by the heart, arthropod blood—hemolymph—flows through openings in the heart located on the dorsal, or upper side, above the gut. The fluid bathes the internal organs.

Insects typically reproduce sexually and lay eggs. However, parthenogenesis, giving birth to unfertilized young, occurs in many insect orders. In some cases, the unfertilized embryo becomes a male; in others, a female; and in still others, either sex can occur. Parthenogenesis is particularly common among the Hymenoptera (bees and wasps). A few insects give birth to live young.

Insect Morphology

Morphology is the study of the physical structure of an organism. To correctly identify insects, one must understand basic insect morphology. Insects have three body parts: head, thorax, and abdomen. The head region contains the mouthparts, eyes, and antennae. The thorax region includes the legs and wings, and the abdomen contains the anus and reproductive structures.

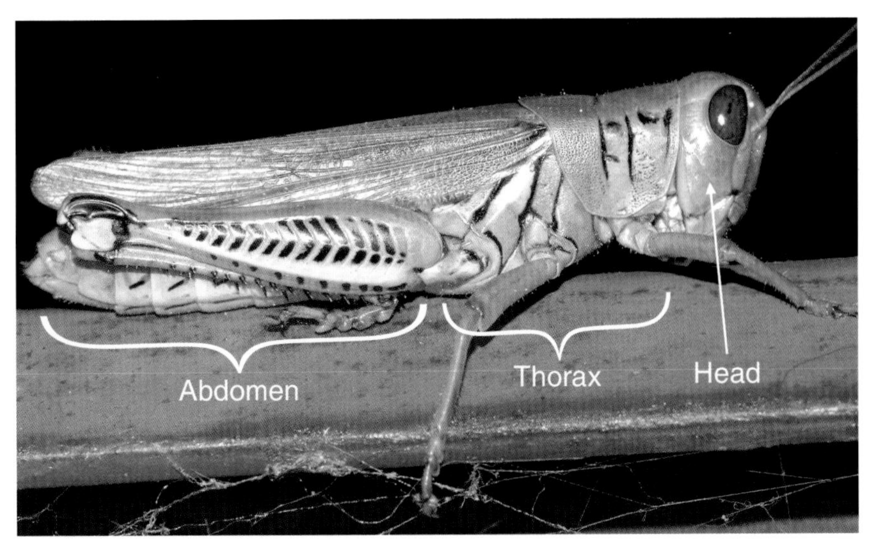

Body regions

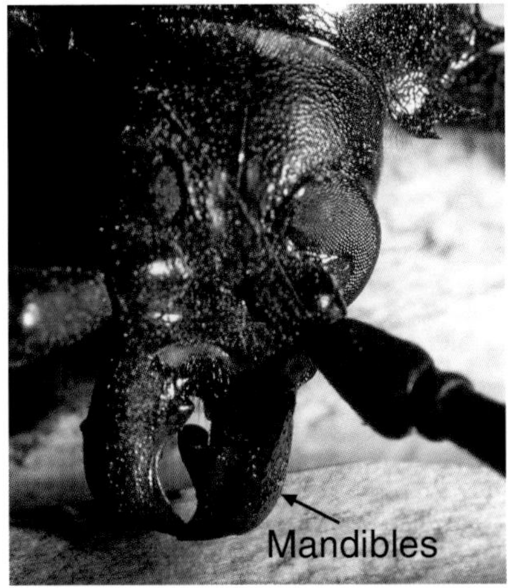

Chewing mouthparts

Insect Mouthparts

Insect mouthparts vary considerably depending upon how the insect feeds. Beetles, grasshoppers, cockroaches, mantids, and others have chewing mouthparts with distinct toothlike structures called mandibles. In addition to having mandibles, these insects have fingerlike appendages called palps that are used in the feeding process.

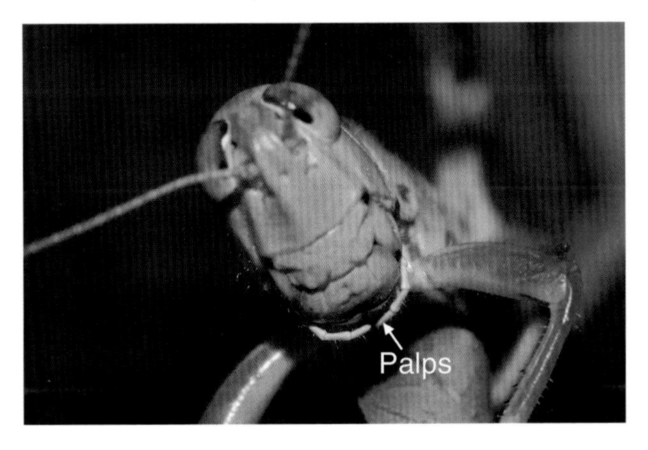

Palps

Mosquitoes, planthoppers, true bugs, and others have mouthparts adapted for piercing and sucking fluids from their hosts. These insects often inject enzymes into the host that dissolve and liquefy the surrounding tissue. The swelling and itching of mosquito bites are often our body's reaction to these foreign chemicals. Butterflies and moths have long, tubelike mouthparts designed to reach deep into flowers to siphon nectar.

Piercing/sucking mouthparts, mosquito

Other insects have mouthparts adapted for their specific needs. House flies have a padlike structure at the tip of their mouthparts that is used to sponge liquefied materials, whereas honey bees have a tonguelike mouthpart used to lap nectar.

Siphoning mouthparts, sulphur butterfly

Sponging mouthparts, house fly

Piercing/sucking mouthparts, assassin bug

Siphoning mouthparts, satyr butterfly

Lapping mouthparts, honey bee

Insect Antennae

Antennae are sensory organs that allow insects to detect food sources, find mates, locate colonies, and perform other important sensory functions. The antennal type is often an important identification tool.

Filiform antennae, the most common antennal type, are long and threadlike. Other distinctive antennal forms are flag shaped (lamellate), elbowed (geniculate), sawlike (serrate), feathery (plumose), and clubbed (clavate).

Filiform antennae, cricket

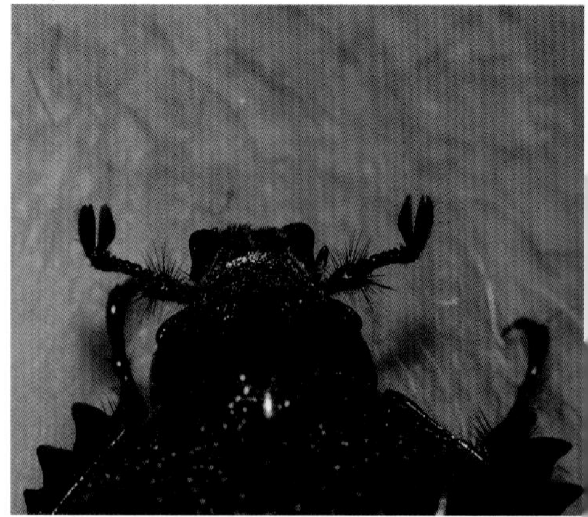

Lamellate antennae, scarab beetle

Filiform antennae, cockroach

Serrate antennae, net-winged beetle

Geniculate antennae, ant

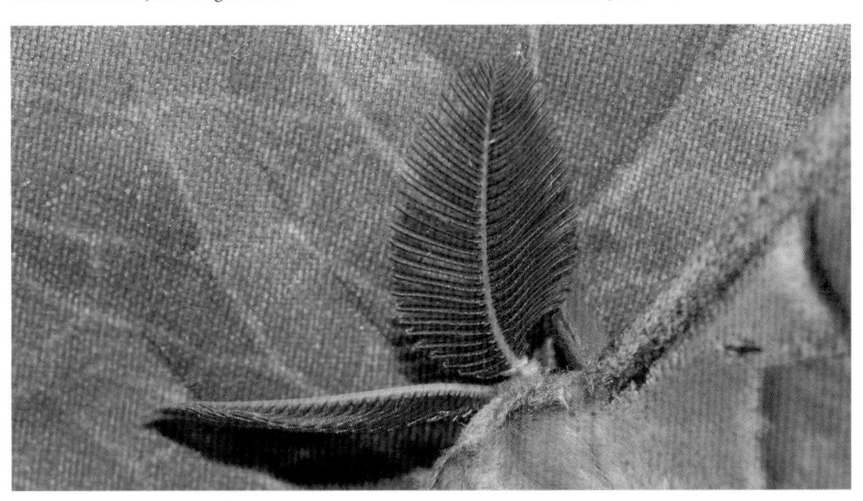

Plumose antennae, polyphemus moth

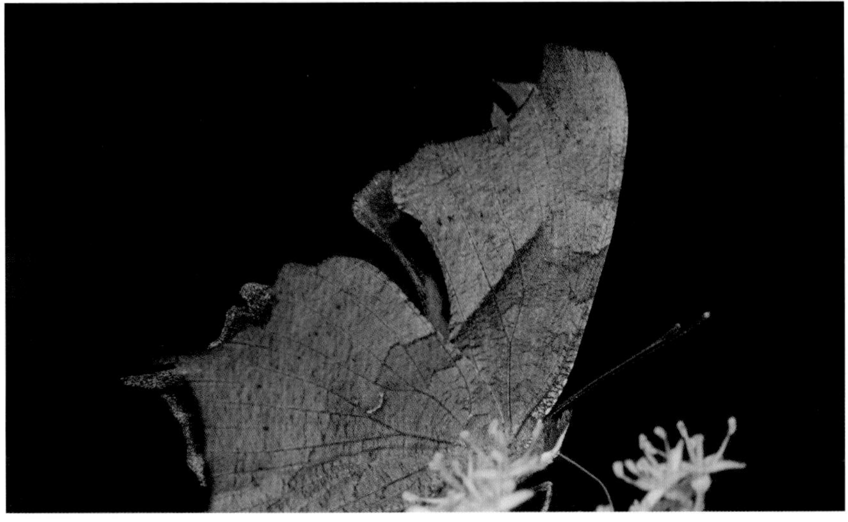

Clavate antennae, question mark butterfly

Grasshopper thorax

Odonata thorax

Insect Thorax

The middle section of an insect body is the thorax. From the head rearward, the thorax is divided into three segments: prothorax, mesothorax, and metathorax. Each thoracic segment has one pair of legs, but only the meso- and metathoracic segments possess wings.

The prothorax is often an important morphological trait. The top portion of the prothorax, called the pronotum, is elongated and covers the head of cockroaches (order Blattodea), fireflies (order Coleoptera), and a few other groups. Treehoppers (order Hemiptera, suborder Auchenorrhyncha) are often readily recognized by their ornate and often bizarre-shaped pronotum.

Adult with fully developed wings

Insect Wings

Insects are the only arthropods that have wings, yet not all insects possess them. Many insects, such as springtails (order Collembola), silverfish (order Thysanura), and cave crickets (order Orthoptera), do not have wings and therefore have lifestyles that are not adapted to flying. In some insect species, such as some ants and termites, only the reproductive queens and kings have wings. The nonreproductive workers typically live in the soil and never develop nor have the need for wings.

Although most flying insects have

Nymph with wing buds

two pairs of wings, the true flies (order Diptera) have one fully developed pair on the mesothorax and a second reduced pair, called halteres, on the metathoracic segment.

Only adult insects have fully developed and functional wings. The immature stages of some orders, such as Orthoptera and Hemiptera, have small wing pads, but these stages are not capable of flight. Students often bring me small flies, beetles, or other insects and ask if they are baby insects because of their size. Remember, regardless of size, only adults have fully developed wings; although a small insect may resemble a large one, it is probably a different species.

Insect Legs

Most insects have one pair of legs on each of the three thoracic segments. However, some species, such as scales, are adapted to a sedentary lifestyle and possess legs only during the early stages of development.

Most insects have legs designed for walking. Some insects, however, have modified legs that are designed for a specific function. The mantid has grasping forelegs well suited for its predaceous lifestyle. Aquatic insects can "walk on water" because they have tiny hairs on their legs that increase the area that contacts the surface of the water.

Grasping forelegs, mantid

Walking legs, tiger beetle

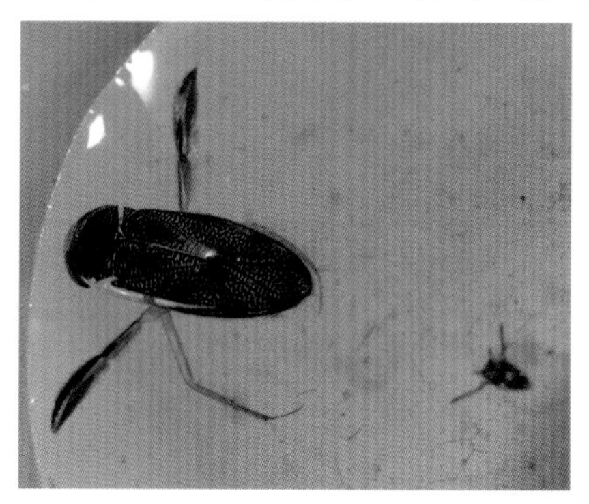

Swimming legs, water boatman

Other insect species have different leg adaptations for their specific lifestyle.

Metamorphosis

Metamorphosis may be defined as the changes that occur as an animal develops from birth to sexually mature adulthood. Insects have an exoskeleton made of a resistant chitinous material that provides protection and serves as a support system to which muscles attach, much as our bones do for us. This external shell typically hardens through a process known as sclerotization.

As the insect develops and grows, the relatively inflexible shell must be replaced at regular intervals. Molting, also referred to as "shedding its skin," occurs at regular intervals that vary according to a particular species. Most species molt four to eight times, but some species molt as many as twenty-eight times. The stage between molts is called an instar.

Insect orders can be separated into four groups based on the type of metamorphosis and the stages of development.

Ametamorphosis

Insects with ametamorphic development pass through three life stages: egg, nymph, and adult. The immature stage resembles the adult, and the only way to distinguish the age of the insect is by its size. Insect orders with ametamorphic development include Collembola (springtails) and Thysanura (silverfish).

Ametamorphosis, springtail

Ametamorphosis, silverfish

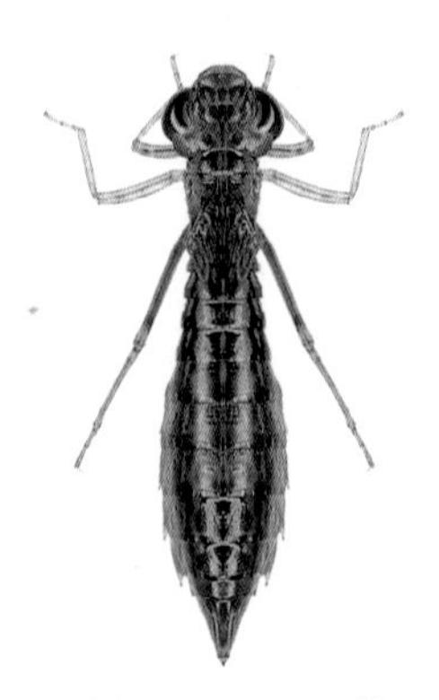

Dragonfly naiad (by James Lasswell)

Incomplete Metamorphosis

Dragonflies and damselflies (Odonata) and mayflies (Ephemeroptera) have incomplete metamorphosis. The stages of development for these insects include the egg, naiad, and adult. The immature naiad is aquatic and possesses gills for breathing underwater.

Gradual Metamorphosis

Insects with gradual metamorphosis pass through an egg stage, two or more nymphal instars, and the adult stage. Nymphs resemble adults but can generally be identified by the lack of fully developed wings. As the nymph develops, wing pads become apparent, but the insect is not capable of flight or reproduction.

Gradual metamorphosis occurs in the following insect orders:

- Phasmatodea (walkingsticks)
- Orthoptera (grasshoppers, crickets, and others)
- Mantodea (mantids)
- Blattodea (cockroaches)
- Isoptera (termites)
- Dermaptera (earwigs)
- Embiidina (webspinners)
- Plecoptera (stoneflies)

Dragonfly adult

Squash bug nymph

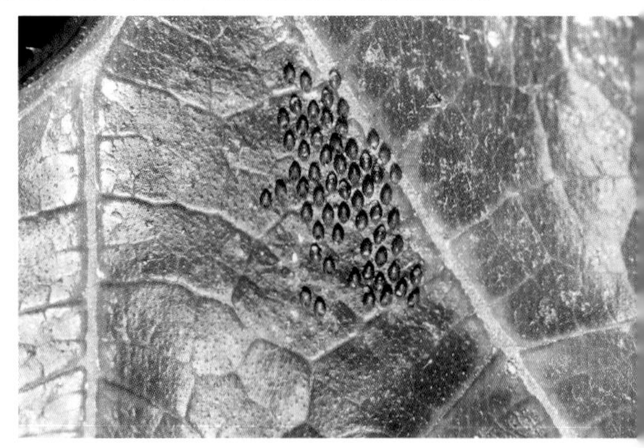
Squash bug eggs

Psocoptera (booklice and barklice)
Mallophaga (biting lice)
Anoplura (sucking lice)
Hemiptera, suborder Heteroptera (true bugs)
Hemiptera, suborders Auchenor rhyncha and Sternorrhyncha (cicadas, hoppers, and others)
Thysanoptera (thrips)

Squash bug adult

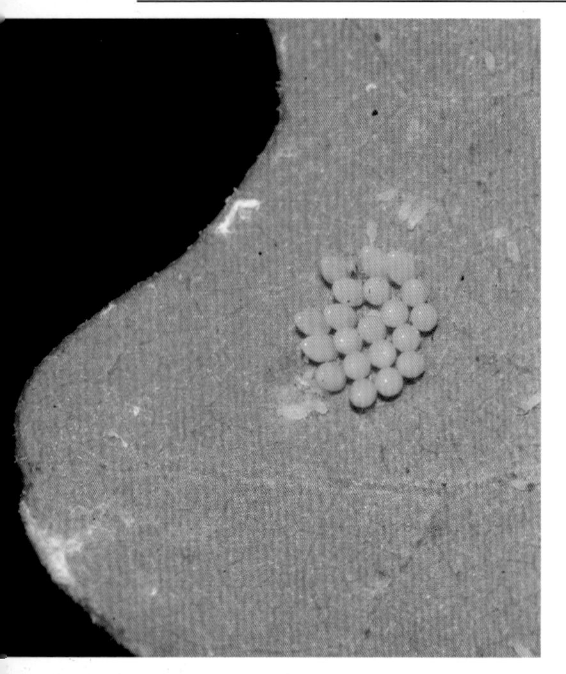

Ladybird beetle larva

Ladybird beetle eggs

Complete Metamorphosis

Insect orders with complete metamorphosis develop through four distinct phases: egg, larva, pupa, and adult. The larval stage typically does not resemble the adult and is not reproductive. Because this stage is the most actively growing and often requires the most food, it is often the most pestiferous or beneficial stage of development.

Following the last larval instar, the insect transforms into a nonfeeding and typically nonmoving intermediate stage called a pupa, often in a prepared cocoon. The final stage of development, the adult stage, possesses fully functional

Ladybird beetle pupa

Ladybird beetle adult

wings and is capable of reproduction. Complete metamorphosis occurs in the following insect orders:

- Neuroptera (dobsonflies, antlions, and others)
- Coleoptera (beetles)
- Mecoptera (scorpionflies)
- Siphonaptera (fleas)
- Diptera (flies)
- Trichoptera (caddisflies)
- Lepidoptera (butterflies and moths)
- Hymenoptera (bees and wasps)

How to Use This Book

This book was designed to help you identify groups of insects using three criteria: picture identification, common physical traits, and the biology and ecology of the group.

First, use the photographs to place the specimen in one of the groups described in the following photo key. In many cases, you will be able to readily accomplish this step. For example, most people can recognize a fly, a butterfly, a beetle, and many other insects. However, some arthropods, particularly the smaller ones, are not as common, and you must determine the order to which they belong.

Once you have established the group, turn to that section, scan the photographs looking for similarities, and review the characteristics listed. The insect families are arranged in alphabetical order according to their common names, so you do not have to be familiar with the family names or the group's evolutionary history.

The physical characteristics of each entry are given, and those that are particularly important in identifying a specimen in the field are in bold print. (A glossary is included with terms that relate to the characteristics described.) Often the biology and/or ecology of the specimen are as important as its physical characteristics in identifying the organism. These key characteristics may include preferred habitats and food sources in addition to the physical traits.

In some cases, the number of antennal and mouthpart segments, as well as the segments in the foot, are important identifying characteristics. These traits are visible only under magnification and are given if a more definitive identification is needed.

The combined physical characteristics, biology, and ecology of the specimen should enable you to place the specimen in the proper group. However, it is important to remember that there are exceptions to almost every rule. If a definitive classification is needed, you should consult other references or a trained entomologist.

Most scientific names are derived from Latin or Greek terms. To assist the reader in pronouncing and remembering the arthropod group's name, the meanings of the root words are given and a pronunciation guide has been developed. The following pronunciation legend can be used:

\ a \ as *a* in *cat*
\ ah \ as *a* in *father*
\ ai \ as *ai* in *stair*
\ ay \ as *ay* in *stay*
\ ee \ as *ea* in *easy*
\ eh \ as *e* in *red*
\ ih \ as *i* in *hit*
\ oh \ as *o* in *go*
\ oo \ as *oo* in *boot*
\ oy \ as *oy* in *boy*
\ uh \ as *u* in *up*
\ y \ as *i* in *lime*

\ b \ as *b* in *boot*
\ d \ as *d* in *day*
\ f \ as *f* in *finger*
\ g \ as *g* in *go*

\ h \ as *h* in *hot*
\ j \ as *j* in *judge*
\ k \ as *k* in *kite*
\ l \ as *l* in *like*
\ m \ as *m* in *moon*
\ n \ as *n* in *not*
\ p \ as *p* in *pipe*
\ r \ as *r* in *right*
\ s \ as *s* in *sat*
\ t \ as *t* in *tap*
\ v \ as *v* in *vivid*
\ z \ as *z* in *zoo*

The bulk of this book describes the orders and families of insects that are commonly found in Texas. Following are the common arthropods and the characteristics most commonly associated with that group.

Common Name	*Class*	*Order*	*Page*
Arachnids	Arachnida		32
Spiders	"	Araneae	33
Scorpions	"	Scorpiones	34
Whipscorpions	"	Uropygi	35
Harvestmen	"	Opiliones	36
Mites and ticks	"	Acari	37
Pseudoscorpions	"	Pseudoscorpiones	38
Windscorpions	"	Solifugae	39
Crustaceans	Crustacea		40
Sowbugs and pillbugs	"	Isopoda	40
Crayfish and crabs	"	Decapoda	40
Millipedes	Diplopoda	(several)	41
Centipedes	Chilopoda	(several)	42
Insects	Hexapoda		43
Springtails	"	Collembola	44
Silverfish	"	Thysanura	45
Mayflies	"	Ephemeroptera	46
Dragonflies and damselflies	"	Odonata	47
Walkingsticks	"	Phasmatodea	54
Grasshoppers, katydids, crickets	"	Orthoptera	55
Mantids	"	Mantodea	62
Cockroaches	"	Blattodea	63
Termites	"	Isoptera	64
Earwigs	"	Dermaptera	65
Webspinners	"	Embiidina	66
Stoneflies	"	Plecoptera	67

Booklice and barklice	"	Psocoptera	68
Chewing lice	"	Phthiraptera (suborder Mallophaga)	69
Sucking lice	"	Phthiraptera (suborder Anoplura)	70
True bugs	"	Hemiptera (suborder Heteroptera)	71
Aphids, cicadas, hoppers	"	Hemiptera (suborders Auchenorrhyncha and Sternorrhyncha)	92
Thrips	"	Thysanoptera	101
Antlions, dobsonflies, lacewings	"	Neuroptera	102
Beetles	"	Coleoptera	110
Scorpionflies	"	Mecoptera	135
Fleas	"	Siphonaptera	136
Flies	"	Diptera	137
Caddisflies	"	Trichoptera	157
Butterflies, moths, skippers	"	Lepidoptera	158
Wasps, ants, bees	"	Hymenoptera	180

Arachnids, Class Arachnida

Four pairs of legs
Two body regions (sometimes indistinguishable)
No antennae

Spiders

Order Araneae
Two distinct body regions
Often produce silk

Spider

page 33–39

Scorpions

Order Scorpiones
Abdomen ends in a tail and possesses a stinger
Pincherlike pedipalps

Scorpion

page 34

Whipscorpions

Order Uropygi
Large pincherlike pedipalps
Long whiplike tail
Typically found in arid regions

Whipscorpion

page 35

Harvestmen

Order Opiliones
Appear to have only one body region
Long, slender legs
Often found in groups

Harvestman

page 36

Mites, Ticks

Order Acari
Appear to have only one body region
Legs short
Some parasitic on mammals; some feed on plants; some predaceous on other arthropods

Tick

page 37

Pseudoscorpions

Order Pseudoscorpiones
Very small
Resemble scorpions but do not have a tail or stinger
Typically found in plant litter

Pseudoscorpion

page 38

Windscorpions
Order Solifugae
Cream colored
Large, dark jaws
Most commonly occur in arid regions

Windscorpion (by John Jackman)

page 39

Crustaceans, Class Crustacea
Body segments variable
Two pairs of antennae, one pair smaller than other pair
Number of legs variable

Crustaceans

page 40

Millipedes, Class Diplopoda
Multisegmented body
Two pairs of legs per body segment
One pair of antennae

Millipede

page 41

Centipedes, Class Chilopoda
Multisegmented body
One pair of legs per body segment
One pair of antennae

Centipede

page 42

Insects, Class Hexapoda
Three body segments
Typically, six legs
Often possess wings
Habitat extremely variable

Insect

page 43–197

Springtails
Order Collembola
Very small
Wingless
Can jump using a springlike forked tail
Common in decomposing leaf litter

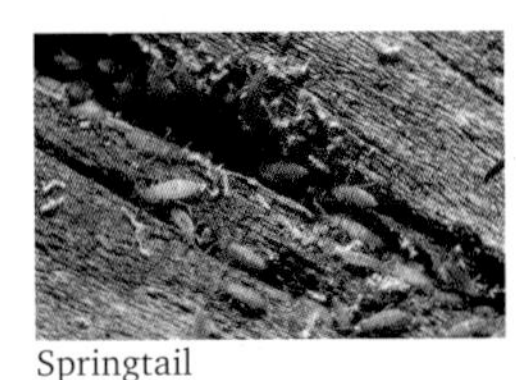
Springtail

page 44

Silverfish
Order Thysanura
Wingless
Silver and scaly
Three tails
Reclusive; prefer moist dark areas

Silverfish

page 45

Mayflies
Order Ephemeroptera

Two pairs of triangular-shaped wings
Two (sometimes three) long tails
Typically found near water

Mayfly

page 46

Dragonflies, Damselflies
Order Odonata

Two pairs of wings; many cross veins
Large head and thorax with slender abdomen
Typically found near water

Dragonfly

Damselfly

page 47

Walkingsticks
Order Phasmatodea

Resemble twigs
Wingless
Typically found on trees and shrubs

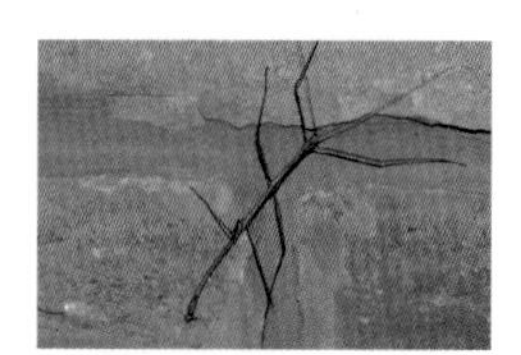
Walkingstick

page 54

Grasshoppers, Katydids, Crickets
Order Orthoptera

Hind legs modified for jumping
Two pairs of wings that lie straight down back; sometimes absent
Typically found on plants or on ground

Short-horned grasshopper

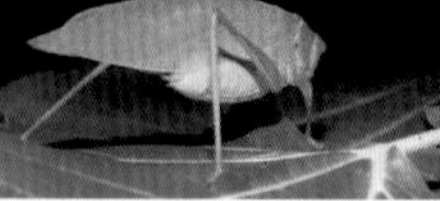
Katydid

Cricket

page 55

Mantids
Order Mantodea

Forelegs modified for grasping
Triangular head that is movable
Two pairs of wings that lie straight down back

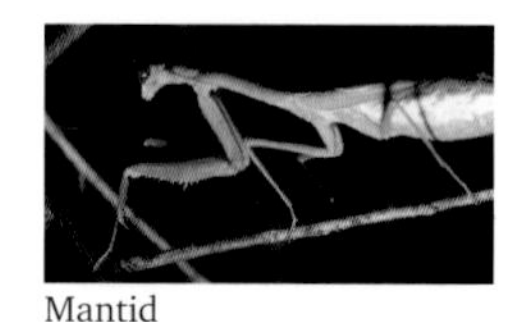
Mantid

page 62

Cockroaches

Order Blattodea

Flattened

Head covered by shield

Two pairs of wings that lie straight down back

Prefer secluded, dark places

Cockroach

page 63

Termites

Order Isoptera

Small, cream colored, sometimes dark

Typically wingless; reproductive class with wings of equal size and shape

Antennae beadlike

Typically found in colonies and associated with wood

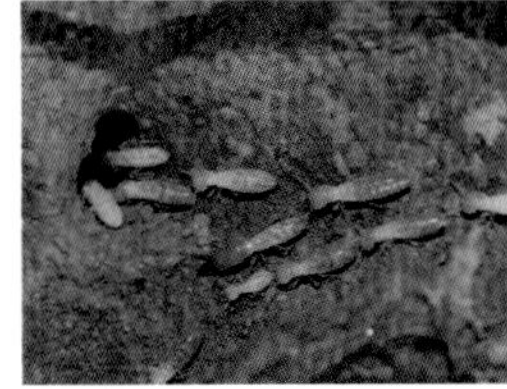
Termites

page 64

Earwigs

Order Dermaptera

Two pairs of wings; first pair short and second folded under first pair

Pincherlike structures

Typically found on ground under protective structures

Earwig

page 65

Webspinners

Order Embiidina

Small

Wingless

Swollen forelegs

Associated with webbing on tree trunks

Webspinner

page 66

Stoneflies

Order Plecoptera

Two pairs of wings that lie flat down back and with diagonal veins

Two short tails

Typically associated with water

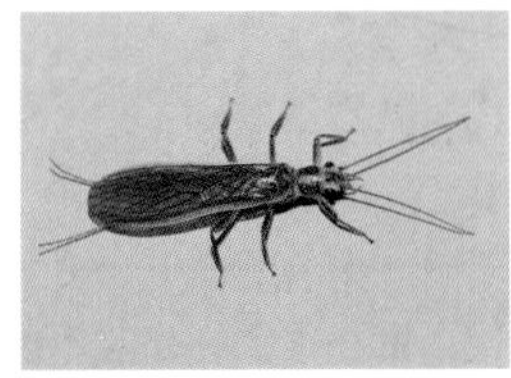
Stonefly

page 67

Booklice, Barklice

Order Psocoptera

Two pairs of wings or wingless

Wings held rooflike

Long antennae

Barklice

page 68

Chewing Lice

Order Phthiraptera

Suborder Mallophaga

Small

Wingless

Head triangular and wider than thorax

Ectoparasites on mammals and birds

Chewing louse

page 69

Sucking Lice

Order Phthiraptera

Suborder Anoplura

Small

Wingless

Head pointed and narrower than thorax

Ectoparasites on mammals

Sucking louse

page 70

True Bugs

Order Hemiptera

Suborder Heteroptera

Two pairs of wings; forewings with two distinct textures

Piercing/sucking mouthparts

Habitats variable

Stink bug

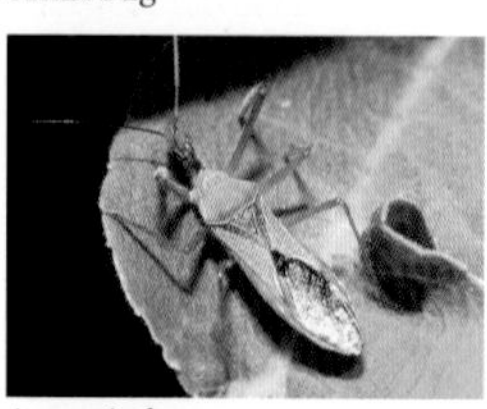
Assassin bug

True bug

page 71

Aphids, Cicadas, Hoppers

Order Hemiptera

Suborders Auchenorrhyncha and Sternorrhyncha

Two pairs of wings; same texture throughout; held rooflike when at rest; a few wingless

Piercing/sucking mouthparts

Found on plants

page 92

Planthopper

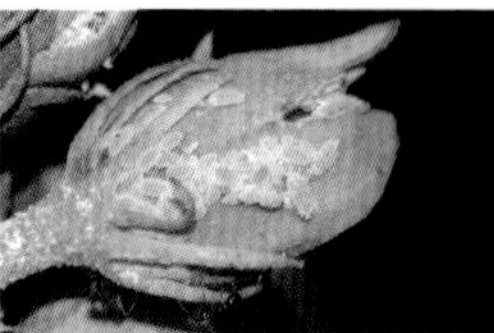
Mealybugs

Aphids

Cicada

Thrips

Order Thysanoptera

Very small

Two pairs of wings; fringed with hairs

Typically associated with plants

page 101

Thrips

Thrips

Antlions, Dobsonflies, Lacewings

Order Neuroptera

Two pairs of wings; many cross veins; held rooflike

Predaceous insects that are typically associated with plants

page 102

Green lacewing

Mantidfly

Beetles

Order Coleoptera

Two pairs of wings; front pair thickened and meet in a straight line when at rest

Chewing mouthparts

Very diverse habitats

Green June beetle

page 110

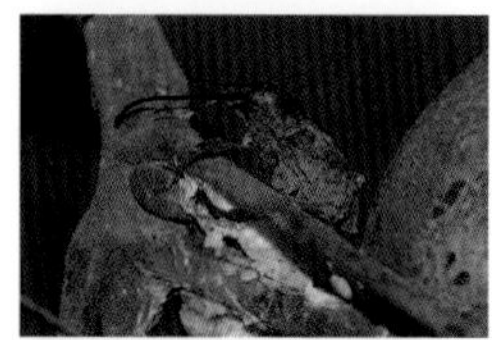
Acorn weevil

Lady beetle

Blister beetle

Scorpionflies

Order Mecoptera

Two pairs of narrow wings

Chewing mouthparts; located at the end of an elongated snout

Some males' genitalia resemble scorpion stinger

Scorpionfly

page 135

Fleas

Order Siphonaptera

Wingless

Laterally compressed

Ectoparasites on mammals

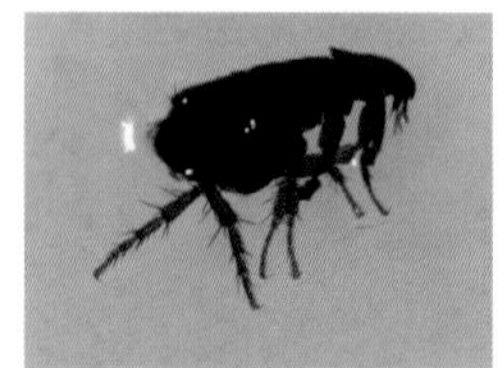
Flea

page 136

Flies

Order Diptera

One pair of wings

Mouthparts variable

Habitat variable

page 137

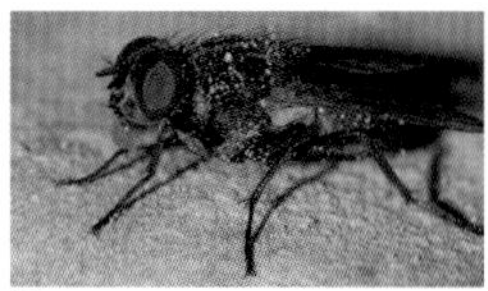

Fly

Crane fly

Robber fly

Mosquito

Caddisflies

Order Trichoptera

Two pairs of wings; typically with short hairs or scales

Resemble small moths

Long antennae

Typically associated with water

page 157

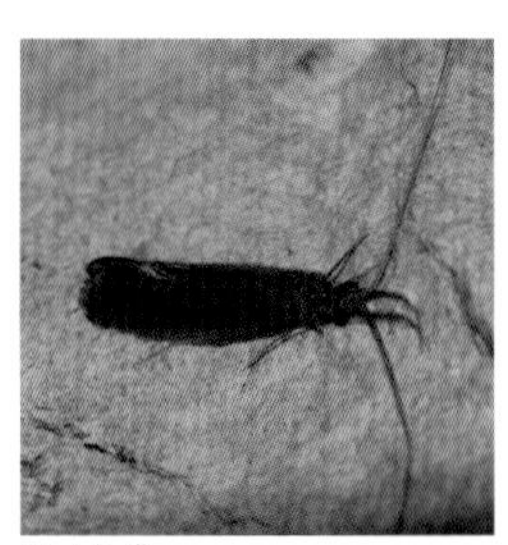

Caddisfly

Butterflies, Moths, Skippers

Order Lepidoptera

Two pairs of wings; covered with scales
Mouthparts long and coiled
Typically associated with plants

page 158

Buckeye butterfly

Polyphemus moth

Skipper

Wasps, Ants, Bees

Order Hymenoptera

Typically two pairs of wings; sometimes absent; generally transparent
Some possess a sting
Most have a restriction between the thorax and abdomen
Often found in colonies
Habitat variable

page 180

Ichneumon wasp

Paper wasp

Honey bee

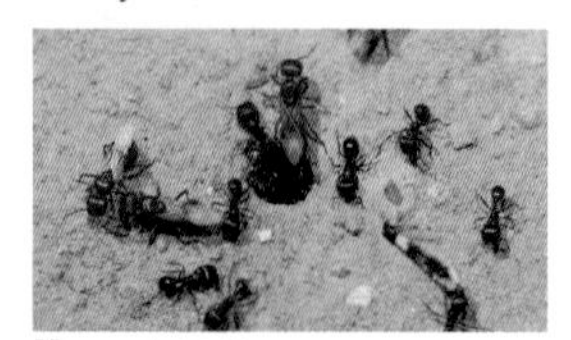
Harvester ant

Arthropods

PHYLUM: ARTHROPODA
(ar-THRAH-pih-duh)
arthro, jointed; *poda*, foot

Characteristics

SIZE: Tiny to >4 in. (10 cm)
ANTENNAE: None to two pairs
MOUTHPARTS: Variable
EYES: Compound; usually present
WINGS: Present or absent
LEGS: Variable; jointed
MISCELLANEOUS: **Possess an external skeletal system called an exoskeleton**
HABITAT: Extremely variable
FOOD: Extremely variable

Arthropoda is the largest and most diverse phylum in the animal kingdom with an estimated 6 million species worldwide. These generally small creatures have a hard exoskeleton made of chitin that does not expand as they grow but must be periodically shed and replaced through a process known as molting. Arthropods also have paired, jointed appendages, including legs, antennae, and fingerlike projections, called pedipalps, near the mouth.

The more common arthropods found in Texas include members of the arachnids, crustaceans, millipedes, centipedes, and insects. Their biological and ecological diversity is unmatched by any other group of organisms, and they consume an extremely diverse array of foods.

CLASSES IN THE PHYLUM ARTHROPODA

Class	*Page*
Arachnids: Arachnida	32–39
Crustaceans: Crustacea	40
Millipedes: Diplopoda	41
Centipedes: Chilopoda	42
Insects: Hexapoda	43–197

Arachnids

Class: Arachnida
(uh-RAK-nih-duh)
arachne, spider

Characteristics:

SIZE: Variable; very tiny to >4 in. (10 cm)
SHAPE: **Two body segments, sometimes indistinguishable**
ANTENNAE: Absent
Mouthparts: Possess chelicerae and pedipalps; pedipalps often modified as pinchers
EYES: Usually present
WINGS: Absent
LEGS: **Four pairs**
HABITAT: Mostly terrestrial
FOOD: Most predaceous on other arthropods

This large group includes the very common and well-known spiders, scorpions, harvestmen, mites, and ticks. Some less common and infrequently observed members are the large whipscorpions, the very tiny pseudoscorpions, and the windscorpions.

This class of arthropods is recognized by having a two-segmented body. The head and thorax are combined to form a cephalothorax to which the legs are attached. The abdomen is distinctly separated from the cephalothorax in some groups (spiders) but is indistinguishable in others (mites and ticks, harvestmen, and others). All members of this class possess four pairs of legs as adults.

Arachnids have specialized mouthparts composed of the chelicerae and pedipalps. The chelicerae hold the prey and serve as the fangs in spiders or as jaws in other members of the class. The pedipalps are often fingerlike (spiders, ticks, and mites). However, the pedipalps can be distinctly modified to form pinchers (scorpions), or they may be long and leglike (windscorpions).

Some members, particularly the spiders, are venomous and secrete their toxin from poison glands through their chelicerae. The scorpions, however, immobilize their prey with a stinger located at the end of their abdomen. Most species are predaceous on other arthropods and are considered beneficial.

COMMON ORDERS IN THE CLASS ARACHNIDA

Order	*Page*
Spiders: Araneae	33
Scorpions: Scorpiones	34
Whipscorpions: Uropygi	35
Harvestmen: Opiliones	36
Mites and ticks: Acari	37
Pseudoscorpions: Pseudoscorpiones	38
Windscorpions: Solifugae	39

Spiders

Order: Araneae
(uh-RAY-nee-ee)

Characteristics:
SIZE: $\frac{1}{8}$ to >4 in. (3.2 mm to 10 cm)
SHAPE: **Two distinct segments: cephalothorax and abdomen**
ANTENNAE: Absent
MOUTHPARTS: Piercing/sucking; **fingerlike pedipalps**
EYES: Present; six to eight
WINGS: Absent
LEGS: **Four pairs**
MISCELLANEOUS: **Abdomen not segmented**
HABITAT: Mostly terrestrial; variable
FOOD: Predaceous on other arthropods

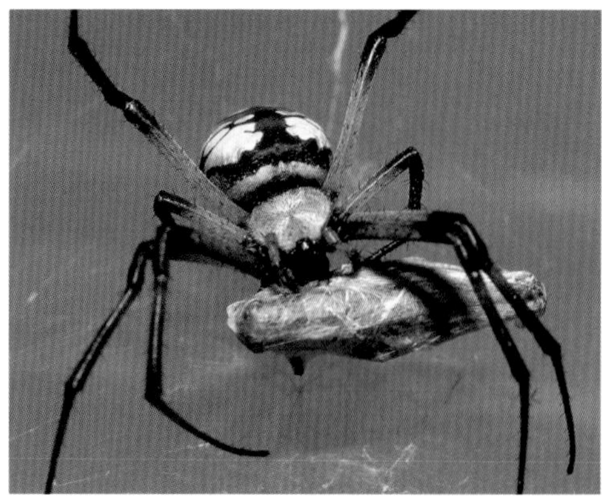

Garden spider with prey

Spiders are easily recognized and generally unwelcome in human habitats. They can be distinguished from other arthropods by their two distinct body segments, four pairs of legs, and fingerlike pedipalps.

Spiders are well known for their silk, which is used in a variety of ways. Orb spiders make elaborate aerial webs to capture prey. Once prey is entangled in the web, the spider will often spray more silk onto the struggling victim to subdue it and finally wrap it in silk to keep the meal fresh. Other spiders produce silk parachutes used to catch puffs of air to carry them to new areas.

Female spiders lay their eggs in silken sacs and typically guard them, at least until the eggs hatch. The young spiders, called spiderlings, resemble the adults but are smaller and may stay with the female for a short time.

All spiders are predaceous, feeding on other arthropods, and are considered beneficial. Although all spiders produce venom, only a very few species are considered dangerous to humans. In Texas, the black widow, *Latrodectus mactans*, and the brown recluse, *Loxoceles reclusa*, are common poisonous spiders.

Brown recluse, *Loxosceles reclusa*

Black widow, *Latrodectus mactans*

Scorpions

Order: Scorpiones
(skor-pee-AH-nees)

Characteristics:
SIZE: Up to 2 ½ in. (6.4 cm)
SHAPE: **Abdomen broadly joined to cephalothorax; last five segments narrow and tail-like**
ANTENNAE: Absent
MOUTHPARTS: Piercing/sucking; **enlarged pincherlike pedipalps**
EYES: Present
WINGS: Absent
LEGS: Four pairs
MISCELLANEOUS: Abdomen segmented
HABITAT: **Arid regions**
FOOD: Predaceous on insects and spiders

Scorpions are easily recognized by their enlarged, pincherlike pedipalps and segmented abdomen that narrows into a tail that ends with a sting. Scorpions have a pair of eyes on top and in front of the cephalothorax and additional pairs on the lateral edges. Their vision, however, is poor, and they mainly rely on other senses.

Scorpions are nocturnal predators that are most common in arid regions. During the day, they hide under rocks and other shelters on the ground. At night, they use comblike structures called pectines located under their abdomen to navigate and sense the presence of other organisms.

Scorpions use their venomous sting to subdue prey and to defend themselves. The sting of most Texas scorpions is comparable to that of a honey bee or paper wasp and not considered life threatening.

Scorpions typically live from three to eight years.

Scorpion

Scorpion mouthparts

Scorpion pectines

Whipscorpions, Vinegaroons

Order: Uropygi (yur-ah-PEE-jee)

Relative size of whipscorpion

Characteristics:

SIZE: **Up to 6 in. (15.2 cm) (including tail)**

SHAPE: **Two distinct body regions: cephalothorax and abdomen**

ANTENNAE: Absent

MOUTHPARTS: Chelicerae; **enlarged, pincherlike pedipalps**

EYES: Present

WINGS: Absent

LEGS: Four pairs; front pair very long and slender

MISCELLANEOUS: **Long whiplike tail; abdomen segmented**

HABITAT: Ground

FOOD: Predaceous on other arthropods

Whipscorpions, also known as vinegaroons, are large, fierce-looking creatures found mostly in the arid regions of West Texas. They are easily recognized by their large, stout pedipalps that resemble pinchers and their long, whiplike tail. Their forelegs are longer and more slender than the other legs and may resemble antennae. The only species in Texas, *Mastigoproctus giganteus*, can reach over three inches long excluding the tail!

Whipscorpions are nocturnal predators that use their enlarged pedipalps to capture other arthropods living on the ground. The name "vinegaroon" stems from their ability to spray a mist of concentrated acetic acid (the same chemical found in vinegar) from glands located at the base of their tail.

A female whipscorpion may require several years to reach sexual maturity. After laying her eggs in secluded places, she protects them from predators, and the young whipscorpions remain with their mother through several molts. The adult female dies shortly after her offspring have dispersed.

Whipscorpion

Harvestmen, Daddy-longlegs

Order: Opiliones
(oh-pihl-ee-OH-nees)

Characteristics:

SIZE: Body, $\frac{3}{8}$ in. (9.6 mm)
SHAPE: **Body round; appears to be one segment**
ANTENNAE: Absent
MOUTHPARTS: Chelicerae; **fingerlike pedipalps**
EYES: Usually two present
WINGS: Absent
LEGS: **Four pairs; very long and slender**
MISCELLANEOUS: **Abdomen segmented**
HABITAT: Sheltered areas; under rocks; in caves
FOOD: Predaceous on other arthropods or scavengers on dead insects

Harvestmen resemble and are sometimes mistaken for spiders. The cephalothorax and abdomen of harvestmen are broadly joined and appear to be one unit. Spiders, in contrast, have a distinctly separate cephalothorax and abdomen. Most species of harvestmen have extremely long, slender legs that are used as sensory organs. The legs of some species, however, are not much longer than a typical spider's legs.

Harvestmen are generally found in wooded areas, under rocks, in caves, or in other secluded areas and sometimes in clusters of several hundred individuals. Most species are predaceous on other arthropods or feed on decomposing materials.

Contrary to popular belief, harvestmen are not poisonous and are quite harmless. They do, however, have glands on either side of their abdomen that can release a rather disagreeable odor when the harvestman is disturbed.

The name "daddy-longlegs" applies only to those harvestmen in the family Phalangidae, which are common in Texas.

Harvestman

Harvestman, side view

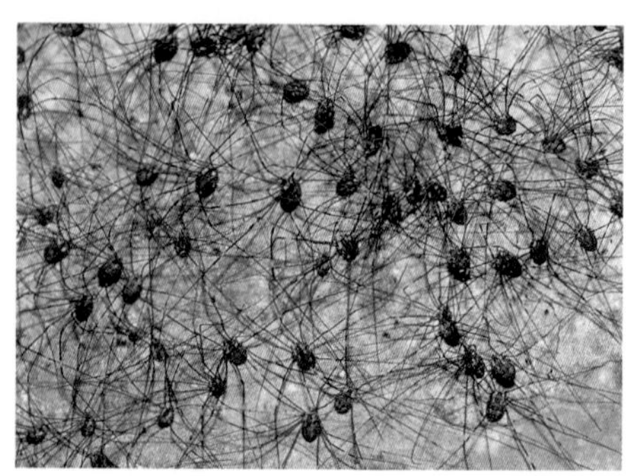

Harvestmen cluster

Mites, Ticks

Order: Acari
(AK-uh-ree)

Characteristics:

SIZE: **Very tiny to $\frac{3}{8}$ in. (9.6 mm)**
SHAPE: **Oval; appear to have one body segment**
ANTENNAE: Absent
MOUTHPARTS: Piercing/sucking
EYES: Variable
WINGS: Absent
LEGS: **Three pairs in larval stage; four pairs in nymph and adult stages**
MISCELLANEOUS: Abdomen not segmented
HABITAT: Variable
FOOD: Variable: plant feeders; parasitic on animals; a few predaceous on other small arthropods

Mites and ticks are distinguishable from other arthropods by their oval body, unsegmented abdomen, and relatively short legs. They have four life stages: egg, six-legged larva, eight-legged nymph, and adult.

Ticks are blood-sucking parasites of animals including mammals, birds, and reptiles. Ticks can be separated into two families: the hard ticks (Ixodidae) and the soft ticks (Argasidae). Hard ticks have a hardened dorsal shield, and the mouthparts can be seen from above. Soft ticks lack the dorsal shield, and their mouthparts point downward and cannot be viewed from above.

Mites are much smaller than ticks and often can be seen only with the aid of a magnifying lens. This group of tiny creatures is very diverse in both habitat and food source. Some mites feed exclusively on, and can cause serious damage to, plants. A few species are parasitic, feeding on a variety of animals. Other species, considered beneficial, feed on other small arthropods and arthropod eggs or consume decomposing material.

Mite

Mite webbing (by Gary Brooks)

Tick

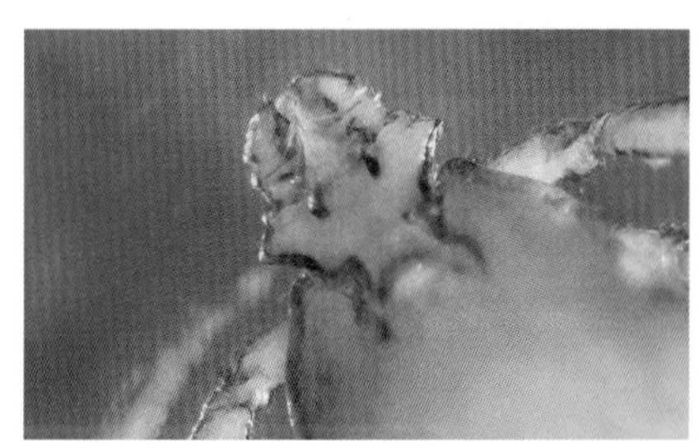

Tick mouthparts

Pseudoscorpions

Order: Pseudoscorpiones
(soo-doh-skor-pee-AH-nees)

Pseudoscorpion

Characteristics:
SIZE: **$\frac{1}{16}$ to $\frac{1}{8}$ in. (1.6 to 3.2 mm)**
SHAPE: **Oval**
ANTENNAE: Absent
MOUTHPARTS: Chelicerae; **large pincherlike pedipalps**
EYES: Variable, none to four
WINGS: Absent
LEGS: Four pairs
HABITAT: **Under bark and in soil debris**
FOOD: Predaceous on small arthropods

As the name implies, pseudoscorpions resemble tiny tail-less scorpions. Their large pedipalps are modified as pinchers, which are used to grasp prey or to catch an occasional ride on a passing fly, beetle, or larger animal.

Pseudoscorpions hide under bark and in debris waiting for an unsuspecting mite, springtail, or other small creature. At the opportune moment, the pseudoscorpion grasps the victim and injects venom from a gland located within its pedipalps.

Pseudoscorpions produce silk, which is used to make a chamber for brooding and overwintering. The female lays her eggs in a silken sac and attaches it to her abdomen. When the eggs hatch, the young remain inside the sac through one molt. Some species can live up to three years.

Pseudoscorpion

Windscorpions

Order: Solifugae
(soh-LIHF-yoo-jee)

Characteristics:

SIZE: $\frac{3}{4}$ to 1 $\frac{1}{2}$ in. (1.9 to 3.8 cm)
SHAPE: **Cylindrical**
ANTENNAE: Absent
MOUTHPARTS: **Chelicerae large and prominent; long, leglike pedipalps**
EYES: Variable
WINGS: Absent
LEGS: Four pairs
MISCELLANEOUS: Generally cream colored; **nocturnal**
HABITAT: **Under rocks during day time; arid regions**
FOOD: **Predaceous on small arthropods**

Windscorpions, also known as sunspiders or camelspiders, are neither scorpions nor spiders but members of the order Solifugae. These arachnids are readily recognized by their large pincherlike chelicerae (fangs) and their long, leglike pedipalps.

Windscorpions prefer warm, arid regions with little or no vegetation. These nocturnal creatures are predators of other small arthropods and have even been reported feeding on small lizards. They are quite rapid runners and are reported to "chase" people but are actually only trying to get into the person's protective shade.

Windscorpions use their long pedipalps to find and seize prey. They then cut the victim into small pieces with powerful chelicerae before sucking the oozing body fluids.

Windscorpions do not possess poison glands but will attempt to bite if handled.

Windscorpion (by John Jackman)

Crustaceans

Class: Crustacea
(kruh-STAY-shuh)
crusta, crust, shell

Characteristics:
SIZE: Variable
SHAPE: **Variable**
ANTENNAE: **Two pairs**
MOUTHPARTS: Chewing
EYES: Usually present
WINGS: Absent
LEGS: **Five or more pairs; first pair sometimes with large claw**
MISCELLANEOUS: Abdomen segmented
HABITAT: **Mostly marine; some fresh water; a few terrestrial**
FOOD: Variable

Pillbugs

Crustacea is a large group of about thirty thousand to fifty thousand species worldwide that live primarily in marine environments. However, several species live in fresh water, and a few are terrestrial. Crustaceans are quite variable in shape and appearance but typically possess two body regions: cephalothorax and abdomen. All crustaceans, however, have two pairs of antennae, although one pair is often smaller.

The order Decapoda includes the familiar lobsters, crayfish, crabs, and shrimp. This group is characterized by having five pairs of legs on the cephalothorax with the first pair modified as large claws. Most members of this group are scavengers.

The order Isopoda is the only terrestrial group of crustaceans and includes the very common sowbugs and pillbugs. Isopods have seven pairs of legs on the cephalothorax. The abdominal region is very short and does not possess legs.

Sowbugs have tail-like projections and cannot roll into a ball, whereas the pillbugs (also known as "roly-polies") do not have these projections and roll up when disturbed. Most isopods are scavengers but occasionally feed on plants and sometimes cause damage.

Sowbugs

Crayfish

Millipedes

Class: Diplopoda
(dihp-LAH-pih-duh)
diplous, double; *poda,* foot

Characteristics:
SIZE: 1 to 4 in. (2.5 to 10 cm)
SHAPE: **Elongated and cylindrical**
ANTENNAE: **One pair; short**
MOUTHPARTS: Chewing
EYES: Present
WINGS: Absent
LEGS: **Two pairs per segment; typically more than thirty pairs**
MISCELLANEOUS: **Nocturnal**
HABITAT: **Damp areas, under soil debris**
FOOD: Mostly scavengers on decomposing plants; some plant feeders; a few predaceous on small arthropods

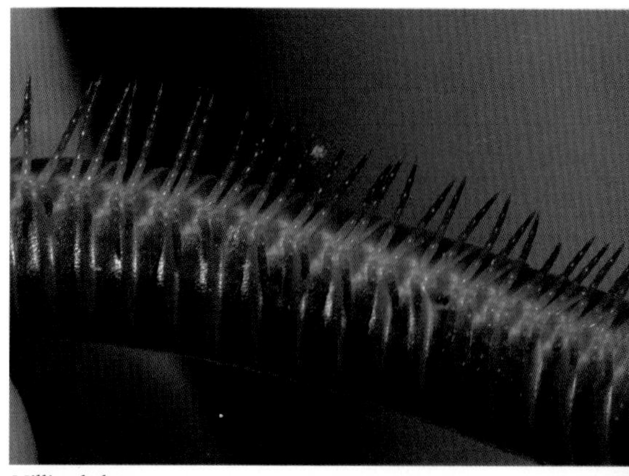

Millipede legs

Millipede

Millipedes live under rocks and debris during the day and venture out at night to scavenge on dead plant materials. These slow-moving creatures are easily recognized by their multisegmented body with two pairs of legs per segment. Most are dark brown and range in size from a little over an inch (2.5 cm) to over four inches (10 cm).

Female millipedes lay their eggs in an earthen cell, which they guard until the young hatch. Young millipedes have only three pairs of legs but add additional legs with each molt. These arthropods overwinter as adults and can live up to seven years.

When disturbed, millipedes roll up into a ball, thus exposing only their hardened exoskeleton. Some species can also emit a cyanide gas that is toxic to most predators.

Centipedes

Class: Chilopoda
(chih-LAH-pih-duh)
centum, hundred; *poda,* foot

Characteristics:
SIZE: Up to 6 in. (15 cm)
SHAPE: **Elongated; flattened**
ANTENNAE: One pair; long
MOUTHPARTS: Chewing
EYES: Present or absent
WINGS: Absent
LEGS: **One pair per segment**
MISCELLANEOUS: **Appendages on first thoracic segment modified as fangs**
HABITAT: **Ground in moist, protected areas**
FOOD: Predaceous on small arthropods

Centipedes are long, slender, flattened arthropods that are generally found outside under ground litter and other damp places. They possess a distinctive head and a multisegmented body with each segment bearing one pair of legs. The number of segments varies, ranging from 15 to 177, and is always an odd number. Most species are a reddish brown. They can live up to six years.

Centipedes are nocturnal and feed on small animals that live under rocks, logs, and soil debris. Fangs with poison glands are located on the first segment and are used to subdue prey and to defend against predators.

The house centipede, *Scutigera coleoptrata,* is well adapted for life in homes. Originally from the Mediterranean region, it was introduced into the United States in the 1800s and has spread across the country. House centipedes prefer moist areas of the house, such as bathrooms, kitchens, and cellars. They hunt for food at night and, when surprised, run rapidly across the floor to a secluded place.

Centipede

Insects

Class: Hexapoda

(hehks-AH-pih-duh)

hexa, six; *poda*, foot

Characteristics:

SIZE: Variable
ANTENNAE: **Present; one pair**
MOUTHPARTS: Variable
EYES: Compound; generally present
WINGS: **Variable; none or one or two pairs**
LEGS: **Generally three pairs, sometimes none**
MISCELLANEOUS: **3-segmented body**
HABITAT: Extremely variable
FOOD: Extremely variable
METAMORPHOSIS: Variable

Insects are the largest group of living organisms on earth. Estimates range from seven hundred thousand to over 30 million species, most of which have not been described. Insects also inhabit a wider range of environments and dine on a greater variety of food than any other group.

The characteristics separating insects from other arthropods include a three-segmented body, the presence of wings, and three pairs of legs. It should be noted, however, that there are many variations and exceptions to these characteristics.

COMMON ORDERS IN THE CLASS HEXAPODA

Order	*Page*
Springtails: Collembola	44
Silverfish: Thysanura	45
Mayflies: Ephemeroptera	46
Dragonflies and damselflies: Odonata	47–53
Walkingsticks: Phasmatodea	54
Grasshoppers, katydids, crickets: Orthoptera	55–61
Mantids: Mantodea	62
Cockroaches: Blattodea	63
Termites: Isoptera	64
Earwigs: Dermaptera	65
Webspinners: Embiidina	66
Stoneflies: Plecoptera	67
Booklice and barklice: Psocoptera	68
Chewing lice: Phthiraptera (suborder Mallophaga)	69
Sucking lice: Phthiraptera (suborder Anoplura)	70
True bugs: Hemiptera (suborder Heteroptera)	71–91
Aphids, cicadas, hoppers: Hemiptera (suborders Auchenorrhyncha and Sternorrhyncha)	92–100
Thrips: Thysanoptera	101
Antlions, dobsonflies, lacewings: Neuroptera	102–09
Beetles: Coleoptera	110–34
Scorpionflies: Mecoptera	135
Fleas: Siphonaptera	136
Flies: Diptera	137–56
Caddisflies: Trichoptera	157
Butterflies, moths, skippers: Lepidoptera	158–79
Wasps, ants, bees: Hymenoptera	180–97

Springtails

Order: Collembola
(koh-LEHM-boh-luh)
coll, glue; *embolos,* bolt or peg

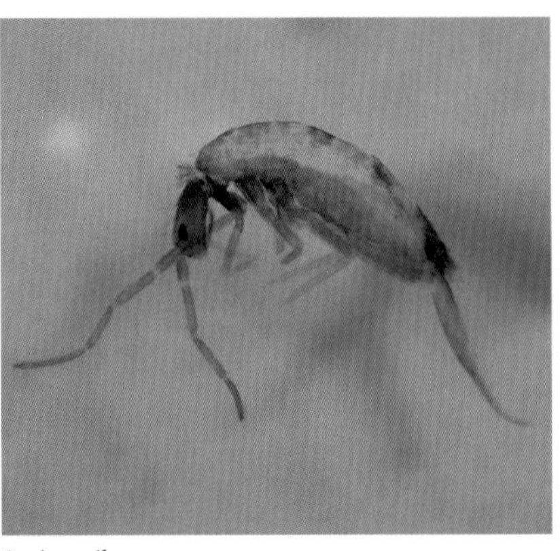

Springtail

Characteristics:

SIZE: **Very small; rarely larger than ⅛ in. (3.2 mm)**
SHAPE: Cylindrical; some species globular
COLOR: Brown, tan, often iridescent
ANTENNAE: Threadlike
MOUTHPARTS: Chewing
EYES: Absent
WINGS: Absent
LEGS: Three pairs
MISCELLANEOUS: Typically hairy; **jumping insects**
HABITAT: Soil and decaying organic matter; prefer moist areas or high humidity
FOOD: Scavengers; mold; algae; pollen
METAMORPHOSIS: Ametamorphous (egg, nymph, adult)

Springtail close-up

Springtails are very small, wingless insects that jump with the aid of a spring-like mechanism.

Springtails often occur in tremendous numbers in moist habitats, where they typically feed on decaying organic matter and microorganisms. In these environments, they are a very important component of the ecosystem, serving both as decomposers and as a food source for predaceous arthropods. Though they are primarily decomposers, they sometimes cause damage to small greenhouse plants, particularly mushrooms when they feed on developing mushroom spores.

The casual observer rarely sees these tiny creatures. However, large numbers commonly may be seen on the sidewalk after a lawn is watered. They can sometimes be seen indoors, for example, on coffee tables, where they wander from houseplant potting soil.

Silverfish, Bristletails, Firebrats

Order: Thysanura
(thy-suh-NYOOR-uh)
thysanos, fringed; *oura*, tail

Characteristics:

SIZE: ⅓ to ½ in. (8.5 to 14.4 mm)
SHAPE: Elongated; flattened
COLOR: **Generally gray or silvery**
ANTENNAE: Long; threadlike
MOUTHPARTS: Chewing
EYES: Compound
WINGS: Absent
LEGS: Walking
MISCELLANEOUS: **Body covered in fine silvery to brown scales; three tails; nocturnal**
HABITAT: Buildings, soil
FOOD: Starches
METAMORPHOSIS: Ametamorphous (egg, nymph, adult)

This group of insects is frequently encountered in homes, where they feed on starchy substances such as wallpaper, linen, and clothing. Although silverfish may simply be nuisances in homes, they can cause serious problems in libraries by eating the glue binding and even the pages of books.

Silverfish prefer cool, damp areas. Firebrats are typically located in warmer areas, such as behind furnaces and water heaters. These insects come out of hiding at night and search for food. They are rapid runners and hide when lights are on. Their shy behavior and speed make them very difficult to catch.

Silverfish

Mayflies

Order: Ephemeroptera
(eh-fehm-uh-RAHP-tur-uh)
ephemera, for a day or short;
ptera, wing

Characteristics:

SIZE: $\frac{1}{3}$ to 1 in. (8.5 to 25.4 mm) (excluding tail)
SHAPE: Cylindrical
COLOR: Typically tan
ANTENNAE: **Short; bristlelike**
MOUTHPARTS: Nonfunctional
EYES: Compound
WINGS: Membranous; **forewings large and triangular; hind wings smaller and rounded**
LEGS: Walking
MISCELLANEOUS: **Two or three long tails;** soft bodied; nocturnal
HABITAT:
Adults: Typically near water
Naiads: Aquatic
FOOD:
Adults: Do not eat
Naiads: Scavengers; vegetation; some predaceous on small aquatic animals

Mayfly

METAMORPHOSIS: Incomplete (egg, naiad, subimago, adult)

After hatching, the aquatic immature mayflies, called naiads, feed on algae and organic matter. When ready to emerge, they rise to the surface of the water in a bubble of air and develop wings in a transitional stage called a subimago. The subimago stage lasts about twenty-four hours before it molts to become an adult.

Adult mayflies do not have functional mouthparts and seldom live more than one or two days. In the spring, male mayflies form large swarms in the air. Attracted to these swirling masses, females fly into the swarm, and they mate. Gravid females lay their eggs on the surface of the water or rocks, or the eggs may fall from a female's dead body after she drops to the surface of the water. The naiads may require one to three years to develop.

Mayflies are very important in aquatic ecosystems. The naiads and dead or dying adults serve as an important food source for many small fish and other aquatic organisms. They are also very selective in the type of aquatic environments where they live, thus serving as indicators of water quality.

Mayfly

Dragonflies, Damselflies

Order: Odonata
(oh-duh-NAH-tuh)
odous, tooth

Characteristics:

SIZE: ¾ to 5 in. (1.9 to 12.7 cm)
SHAPE: **Relatively large head and thorax; slender, elongated abdomen**
COLOR: Variable
ANTENNAE: **Short; bristlelike**
MOUTHPARTS: Chewing
EYES: **Compound; large**
WINGS: Two pairs; many veins
LEGS: Form a basketlike arrangement to catch flying insects
HABITAT:
Adults: Terrestrial; typically near water
Naiads: Aquatic
FOOD:
Adults: Predaceous on other arthropods
Naiads: Predaceous on aquatic arthropods and other small organisms
METAMORPHOSIS: Incomplete (egg, naiad, adult)

This fascinating group of insects is common near almost any source of water. Both the adults and the aquatic naiads are considered beneficial because they consume a wide variety of insects. Being aquatic and sensitive to the chemical changes in their environment, these insects are also good indicators of water quality.

In the field, the Odonata normally can be separated into two distinct suborders. Dragonflies (Anisoptera) are generally large and robust insects, and their eyes touch or are close together. They typically hold their wings flat and outstretched from their body when at rest. Damselflies (Zygoptera), however, are typically smaller and appear more delicate. Their eyes are widely separated, and they normally hold their wings together and above their bodies.

The male's copulating organ is located on the second abdominal segment at the anterior end of the abdomen. When mating, he will clasp the female between the head and thorax with his cerci, which are located at the tip of his abdomen. The female will bend her abdomen down and forward to receive the sperm. After mating, the pair often flies in tandem with the male in front.

Eggs are deposited in vegetation or directly in the water by the female dipping the tip of her abdomen in the water to wash off the eggs. The male typically stays attached to the female during this process while other males often attempt to knock him off and claim the female for their own.

COMMON FAMILIES IN THE ORDER ODONATA

Family	*Page*
Clubtails: Gomphidae	48
Darners: Aeshnidae	49
Skimmers: Libellulidae	50
Broad-winged damselflies: Calopterygidae	51
Narrow-winged damselflies: Coenagrionidae	52
Spread-winged damselflies: Lestidae	53

Clubtails

Family: Gomphidae
(GAHM-fih-dee)

Characteristics:

SIZE: 2 to 3 in. (5.0 to 7.6 cm)
COLOR: **Dark with yellowish or greenish markings**
ANTENNAE: Very short
MOUTHPARTS: Chewing
EYES: Compound; **widely separated**
WINGS: **Clear; no spots**
MISCELLANEOUS: **Often with enlarged terminal portion of abdomen, particularly males**
HABITAT: **Streams and shores of large water bodies**

Clubtail (by James Lasswell)

Clubtail naiad (by James Lasswell)

Clubtail (by James Lasswell)

Clubtail eyes (by James Lasswell)

FOOD:

Adults: Predaceous on other flying insects
Naiads: Predaceous on aquatic insects, other small arthropods, and fish

Clubtails are often recognizable by the swollen abdominal tip, which is typically more prominent in males. Other important characteristics include a dark color with yellow or greenish markings and eyes that do not touch.

A female clubtail generally skims the water, occasionally striking her abdomen on the surface and releasing her eggs. Male clubtails generally do not accompany the female during egg laying. Adult clubtails spend much of their time resting in open sunny areas near water.

The aquatic naiads burrow into the sediment and capture passing prey. When ready to emerge as adults, these naiads crawl onto shore rather than onto aquatic plants.

Darners

Family: Aeshnidae
(ASH-nih-dee)

Darner (by James Lasswell)

Characteristics:

SIZE: 2 to 5 in. (5.0 to 12.7 cm)
COLOR: **Dark brown with bluish or greenish markings**
ANTENNAE: Very short
MOUTHPARTS: Chewing
EYES: Compound; **touch for considerable distance in middle of head**
WINGS: **Always clear**
HABITAT: **Typically calm water; ponds and swamps**
FOOD:
- Adults: Predaceous on other flying insects
- Naiads: Predaceous on aquatic insects, other small arthropods, and fish

Darners are large, powerful dragonflies. The tip of their abdomen is not swollen, and their compound eyes touch in the middle. Darners spend much of their time in flight searching for prey and are often found a considerable distance from water. When resting, they usually land in a vertical position with their abdomen pointed downward.

Darners prefer to breed and develop in calm water. The female repeatedly inserts her needlelike ovipositor into the stem tissue of aquatic plants and lays her eggs. Observers have likened this process to a seamstress darning a sock, thus the common name.

The naiads are active hunters that climb up stems and other underwater structures while searching for prey.

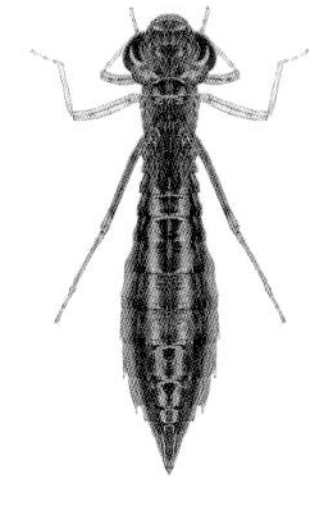

Darner naiad (by James Lasswell)

Darner eyes (by James Lasswell)

Skimmers

Family: Libellulidae
(ly-buh-LOO-lih-dee)

Characteristics:

SIZE: 1 to 3 in. (2.5 to 7.6 cm)

COLOR: **Brightly colored red, yellow, or blue; nonmetallic; female coloration often different from that of mature males**

ANTENNAE: Very short

MOUTHPARTS: Chewing

EYES: Compound; **touch at top of head**

WINGS: **Hind wing with veins near base of wing forming a boot shape with the "toe" of the boot pointing toward the rear edge of the wing; frequently banded or with spots**

MISCELLANEOUS: **Abdomen broad and triangular in cross section; powdery textured**

HABITAT: **Ponds and swamps**

FOOD:

Adults: Predaceous on other flying insects

Naiads: Predaceous on aquatic insects, other small arthropods, and fish

Skimmers, as the name suggests, fly low and erratically over the water. They usually have a very colorful and often powdery-textured body. Many species have spotted or banded wings, and some species have differently colored sexes.

A female skimmer flies close to the water surface; females of some species fly alone, whereas others fly in tandem with a male. The female will occasionally stop, dip her abdomen into the water, and release eggs. The naiads crawl on the floor of the aquatic habitat or climb submerged plant materials but do not burrow into the material.

Male skimmers perch on twigs and other structures in open sunny areas with their wings outstretched and abdomen pointed upward. Skimmer males are quite territorial and will return to the same perch.

Red skimmer

White skimmer (by Joe Carter)

Broad-winged Damselflies

Family: Calopterygidae
(kuh-lahp-tuh-RIHJ-ih-dee)

Characteristics:
SIZE: 2 to 3 in. (5.0 to 7.6 cm)
COLOR: Typically metallic
ANTENNAE: Very short
MOUTHPARTS: Chewing
EYES: Compound; widely separated
WINGS: **Broader than those of other damselflies; held together when at rest; usually colored either completely or at the base**
HABITAT: Streams and rivers
FOOD:
Adults: Predaceous on small, soft-bodied flying insects
Naiads: Predaceous on aquatic insects, other small arthropods, and fish

The calopterygids are the largest damselflies and can often be identified by the striking coloration in their wings. Their wings narrow toward the base rather than form a distinctly stalked base.

Broad-winged damselfly (by James Lasswell)

Broad-winged damselfly (by James Lasswell)

The female broad-winged damselfly lays her eggs in plant tissue, often crawling underwater to reach her preferred site. Males may hover above the submerged female to prevent other males from mating with her.

The naiads are not good swimmers nor active hunters. Immature broadwings are found on underwater vegetation as they wait for food to flow by.

Broad-winged damselfly

Narrow-winged Damselflies

Family: Coenagrionidae
(sihn-ag-ree-AHN-ih-dee)

Characteristics:

SIZE: 1 to 1¼ in. (2.5 to 3.2 cm)
COLOR: Brightly colored; **often blue and black**
ANTENNAE: Very short
MOUTHPARTS: Chewing
EYES: Compound; widely separated
WINGS: **Held together over the back when at rest; clear winged**
HABITAT: **All types of aquatic habitats**

Narrow-winged damselfly (by James Lasswell)

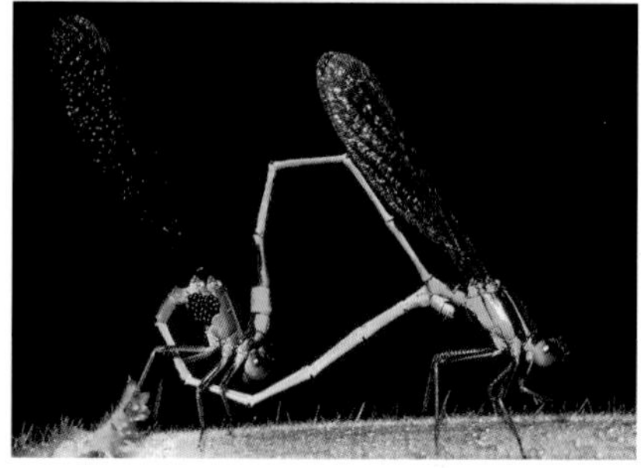
Narrow-winged damselflies mating

Narrow-winged damselfly (by James Lasswell)

Narrow-winged damselfly (by James Lasswell)

FOOD:
- Adults: Predaceous on other flying insects
- Naiads: Predaceous on aquatic insects, other small arthropods, and fish

The coenagrionids are the most common damselflies in Texas. They hold their wings over their back and land in a horizontal position to rest. Sexes are generally different colors with the males typically being brighter than the females. Their wings are clear.

The female narrow-wing uses her ovipositor to make a slit in aquatic plants or wet wood and lays her eggs. She is often accompanied by the male.

The naiads are active hunters that seek small prey among underwater vegetation. The naiads crawl to plants and other materials near the shoreline when they are ready to emerge as adults.

Spread-winged Damselflies

Family: Lestidae
(LEHS-tih-dee)

Characteristics:

SIZE: $1\frac{1}{4}$ to 2 in. (3.2 to 5.0 cm)
COLOR: Bronze, metallic blue, or green
ANTENNAE: Very short
MOUTHPARTS: Chewing
EYES: Compound; widely separated
WINGS: **Often held outstretched when at rest; clear; stalked at base**
HABITAT: **Ponds and protected areas of lakes**
FOOD:
- Adults: Predaceous on other flying insects
- Naiads: Predaceous on aquatic insects, other small arthropods, and fish

Spread-winged damselfly (by James Lasswell)

Spread-winged damselflies are readily recognized by the outstretched wings when first landing to rest. If left unmolested for an extended period of time, they typically fold their wings up over their back. Other damselflies typically hold their wings together and over their abdomen.

Female spread-winged damselflies deposit their eggs, usually while in tandem with males, in aquatic vegetation just above the waterline. They alight in a vertical position in sunny areas to rest.

Spread-winged damselfly (by James Lasswell)

Walkingsticks

Order: Phasmatodea
(faz-muh-TOH-dee-uh)
phasma, ghost

Characteristics:

SIZE: Up to 7 in. (17.8 cm)
SHAPE: Long and cylindrical
COLOR: **Usually resemble the color of tree bark**
ANTENNAE: Long; threadlike
MOUTHPARTS: Chewing
EYES: Compound
WINGS: **Greatly reduced or absent (common)**
LEGS: Walking; long
MISCELLANEOUS: **Resemble sticks or twigs**
HABITAT:
Adults and nymphs: Trees and shrubs
FOOD:
Adults and nymphs: Plants
METAMORPHOSIS: Gradual (egg, nymph, adult)

Walkingstick

Although walkingsticks are plant feeders, they rarely occur in sufficient numbers to cause noticeable damage. In fact, most people are delighted to see them and appreciate their ability to blend in with the environment. In addition to having natural camouflage, these insects produce an offensive odor that serves as a defense mechanism. Even if attacked, walkingsticks have the ability to regenerate lost limbs.

Walkingsticks mating

Some Texas walkingstick species reach over three inches (7.6 cm) in length; some other U.S. species grow over seven inches (17.8 cm) long; and some tropical species can be an incredible one foot (30 cm) long.

There is one generation per year. In the fall, an adult female drops eggs, one at a time, to the ground as she rests in a tree. These eggs, which are the overwintering stage, may hatch the following spring or may require two years. This phenomenon often results in walkingsticks being more abundant in some years than others.

Grasshoppers, Crickets, Katydids, and Others

Order: Orthoptera
(or-THAHP-tur-uh)
ortho, straight; *ptera,* wing

Characteristics:
SIZE: Variable
ANTENNAE: Long; threadlike
MOUTHPARTS: Chewing
EYES: Compound
WINGS: Winged (most common), reduced, or wingless
FOREWINGS: **Held straight down the back when at rest**
HIND WINGS: Broad; folded be neath the forewings when at rest
LEGS: **Hind legs typically modified for jumping**
MISCELLANEOUS: Cerci typically well developed
HABITAT: Plants and soil
FOOD: Many plant feeders; few predaceous on other insects; few scavengers of decomposing matter
METAMORPHOSIS: Gradual (egg, nymph, adult)

One of the more familiar groups of insects, members of the order Orthoptera are generally recognizable by their enlarged rear legs designed for jumping. Their wings lie straight down their back, and they have chewing mouthparts.

Most orthopterans are plant feeders, and several can be serious crop pests. The immature stage typically lives and feeds on the same materials as the adults. In addition, we often are serenaded at night by those species that produce sound by rubbing together certain body parts.

COMMON FAMILIES OF THE ORDER ORTHOPTERA AND THEIR PREFERRED HABITATS

Family	*Soil*	*Under rock, logs, etc.*	*Plants*	*Near water*	*Page*
Camel or cave crickets: Rhaphidophoridae	X	X			56
Crickets: Gryllidae	X	X			57
Katydids, long-horned grasshoppers: Tettigoniidae			X		58
Mole crickets: Gryllotalpidae	X				59
Pygmy grasshoppers: Tetrigidae				X	60
Short-horned grasshoppers: Acrididae			X		61

Camel Crickets, Cave Crickets

Family: Rhaphidophoridae
(raf-ih-doh-FOR-ih-dee)

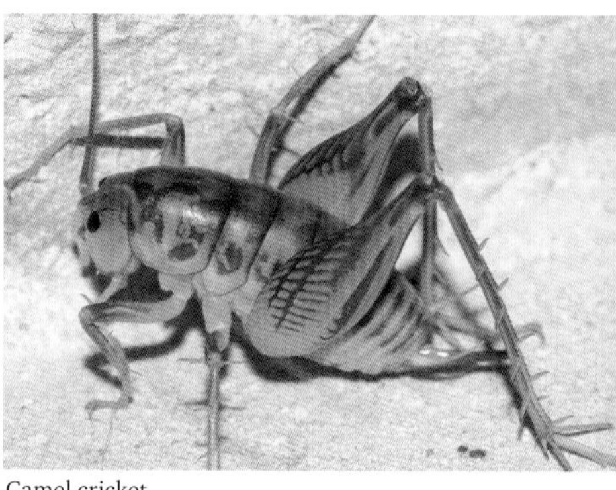
Camel cricket

Characteristics:

SIZE: ¾ to 2 in. (1.9 to 5.0 cm)
SHAPE: Humpbacked
COLOR: **Typically brown**
ANTENNAE: Threadlike; longer than body; touching at bases
MOUTHPARTS: Chewing
EYES: Compound
WINGS: **Wingless**
LEGS: **Hind legs modified for jumping; long spines on tibiae**
MISCELLANEOUS: Nocturnal
HABITAT: **Dark, moist, secluded areas**
FOOD: Decomposing material; some predaceous on small arthropods

Cluster of camel crickets

Both the common names "camel cricket" and "cave cricket" are descriptive of the members of this family. These insects are seldom seen because they spend most of their time in moist areas, such as in caves or under logs, rocks, or any structure that offers protection and is in contact with the ground. They are also known to invade basements and other damp, dark areas of homes.

These wingless insects are easily recognized by their camel-like humped back and long antennae. Their large hind legs possess stout spines and may be used as defensive weapons.

Camel crickets feed mainly on decomposing materials but will eat other smaller insects and arthropods when present.

Crickets

Family: Gryllidae (GRIHL-uh-dee)

Characteristics:

SIZE: $\frac{1}{2}$ to $1\frac{1}{4}$ in. (1.3 to 3.8 cm)
SHAPE: Cylindrical; sometimes flattened
COLOR: Typically dark brown or black; some pale green
ANTENNAE: **Threadlike; longer than body**
MOUTHPARTS: Chewing
EYES: Compound
WINGS: Forewings lie flat over body and **bend sharply over sides**
LEGS: **Hind legs modified for jumping**
MISCELLANEOUS: **Head distinctly rounded; ovipositor spear shaped; two short tails at tip of abdomen; nocturnal**
HABITAT: Hides during daytime
FOOD: Omnivorous

Crickets are well-known insects and one of the more common songsters. Crickets can be separated from other orthopterans by the way their wings bend sharply on the sides and by their rounded heads.

Field crickets, *Gryllus* spp., are probably the most familiar members of the group. They are common in fields and lawns, although they may also enter homes. The house cricket, *Acheta domesticus,* was introduced to the United States from Europe and is common only in the eastern portion of Texas. Tree crickets are typically lime green and seldom seen but are often heard at night as they serenade us from nearby trees and shrubs.

As with other orthopterans, crickets use their wings to produce sound. The sharply bent wings allow these insects to direct the sound to specific areas—often where we are trying to sleep!

Female cricket

Cricket nymph

Tree cricket

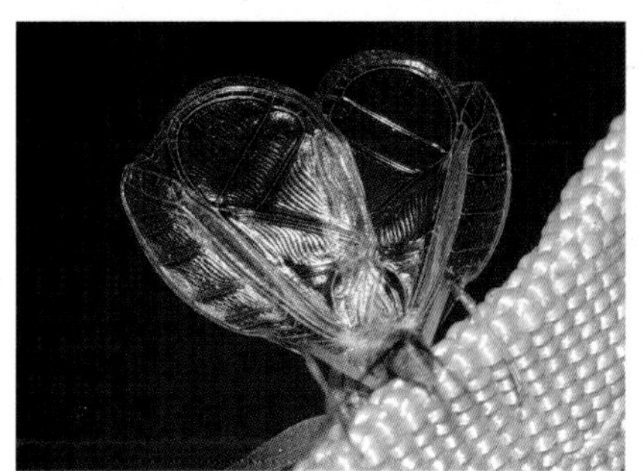
Tree cricket chirping

Katydids, Long-horned Grasshoppers

Family: Tettigoniidae
(teh-tih-guh-NEE-ih-dee)

Characteristics:

SIZE: 1 to 3 in. (2.5 to 7.5 cm)
SHAPE: Variable
COLOR: Generally green or brown; **color often mimics surroundings**
ANTENNAE: **Threadlike; as long as or longer than body**
MOUTHPARTS: Chewing
EYES: Compound
WINGS: Sometimes wingless; **wings sometimes resemble leaves**
LEGS: **Hind legs modified for jumping**
MISCELLANEOUS: **Ovipositor sword shaped**; often possess an ear slit at the base of the front tibiae
HABITAT: Plants
FOOD: Plant feeders; occasionally predaceous on other small arthropods

Long-horned grasshoppers or katydids are less common and cause less damage than short-horned grasshoppers. These insects are typically nocturnal and often difficult to find because many species have wings that look remarkably similar to tree leaves.

Most males can sing by rubbing the edges of their wings together in a process known as stridulating. The female hears the sound by way of an ear slit at the base of her front tibiae. Each species produces a unique sound that is understood by and responded to only by members of its own species.

The female long-horned grasshopper uses her swordlike ovipositor to make a slit in plant tissue or the soil and lay her eggs.

Tettigoniid ovipositor

Katydid

Tettigoniid nymph (by Gary Brooks)

Mole Crickets

Family: Gryllotalpidae
(gril-oh-TAL-pih-dee)

Characteristics:
SIZE: 1 to 2 in. (2.5 to 5.0 cm)
SHAPE: Cylindrical
COLOR: Typically brown
ANTENNAE: Threadlike; shorter than length of the body
MOUTHPARTS: Chewing
EYES: Compound
WINGS: Present; short
LEGS: **Forelegs broad and modified for digging**
MISCELLANEOUS: Hairy; **nocturnal**
HABITAT: **Soil**
FOOD: Predaceous on other insects and small arthropods; plant roots

Mole crickets are nocturnal, soil-inhabiting insects that are seldom seen. These insects are easily recognized by their enlarged forelegs, which are modified for digging. Mole crickets prefer moist, sandy soils or soil that has recently been plowed. They leave their earthen tunnels at night and are capable fliers that are often attracted to lights.

Three species are found in Texas. The southern mole cricket, *Scapteriscus borellii,* was accidentally imported to Georgia in the early 1900s from South America and rapidly spread throughout the warmer regions of the United States. The southern mole cricket feeds primarily on other insects, but its tunneling activities may cause turfgrass roots to dry out and die. The northern mole cricket, *Neocurtilla hexadactyla,* occurs in the eastern portion of Texas and causes damage similar to that of the southern mole cricket.

The tawny mole cricket, *S. vicinus,* is an introduced species that feeds on plant roots and can cause considerable damage to turfgrasses, vegetables, and pastures.

Mole crickets overwinter as adults or large nymphs in earthen cells deep in the soil. In the spring, they emerge, mate, and begin a new generation. Typically, there is one generation per year.

Digging forelegs of a mole cricket

Mole cricket

Mole cricket tunnel (by Gary Brooks)

Pygmy Grasshoppers

Family: Tetrigidae
(teht-RIHJ-ih-dee)

Characteristics:

SIZE: $\frac{3}{8}$ to $\frac{5}{8}$ in. (9.6 to 16.0 mm)
SHAPE: Cylindrical
COLOR: Variable
ANTENNAE: Threadlike; shorter than the length of the body
MOUTHPARTS: Chewing
EYES: Compound; large
WINGS: Present
LEGS: **Hind legs modified for jumping**
MISCELLANEOUS: **Pronotum extends over abdomen and pointed at tip**
HABITAT: **Moist areas**; under leaves in boggy areas or near water
FOOD: Algae and aquatic debris

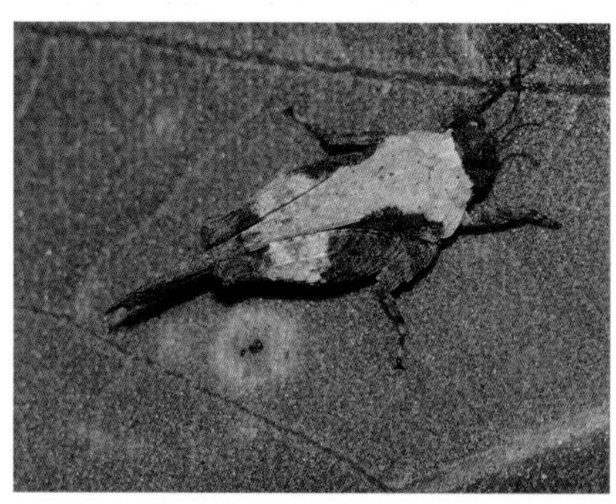

Pygmy grasshopper

Pygmy grasshopper

Pygmy grasshoppers are small orthopterans that resemble short-horned grasshoppers. Under close inspection, what appears to be the top of the wings is actually the extended pronotum, which ends in a point.

Pygmy grasshoppers require moist habitats and can be found under leaves and other plant debris in boggy areas or close to water. These insects are good swimmers and have adapted to their semiaquatic environment by cryptic coloration. Many species exhibit polyornatism, in which several distinct color patterns occur within a population. Pygmy grasshoppers often live longer than a year and spend the winter months as adults or late instar nymphs. Most other grasshoppers overwinter in the egg stage.

Short-horned Grasshoppers

Family: Acrididae
(uh-KRIH-dih-dee)

Characteristics:
SIZE: $\frac{5}{8}$ to $1\frac{1}{4}$ in. (1.6 to 3.0 cm)
SHAPE: Cylindrical
COLOR: Variable
ANTENNAE: Threadlike; **shorter than the length of the body**
MOUTHPARTS: Chewing
EYES: Compound; large
WINGS: Usually present; may be reduced
LEGS: **Hind legs modified for jumping**
MISCELLANEOUS: **Ovipositor short**
HABITAT: Plants
FOOD: Plant feeders

Short-horned grasshopper nymph

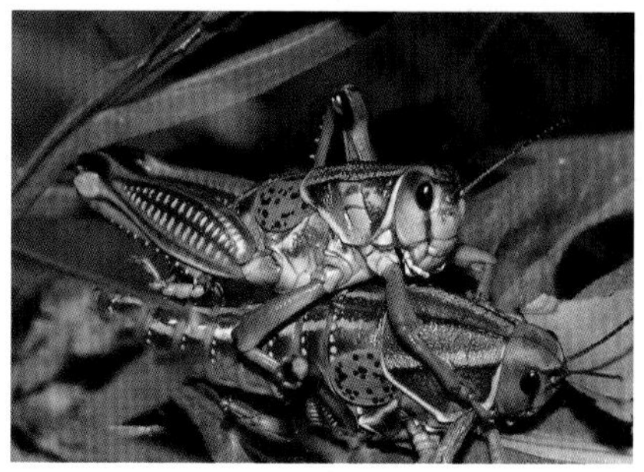
Lubber grasshoppers mating

One of the more infamous groups of insects, short-horned grasshoppers feed on a wide range of plants and can devastate large areas of pasture and other crops. These grasshoppers typically spend the winter as eggs in the soil and emerge in early spring.

The males of many species attempt to attract a mate during the daytime by rubbing their hind femur against the lower edge of the forewing, creating those familiar summer noises. They can also make sound by snapping their wings together during flight. This is believed to frighten birds that may be looking for a flying meal.

Short-horned grasshopper

The differential grasshopper, *Melanoplus differentialis,* is a nonmigratory grasshopper that can cause extensive damage to pastures. It can be recognized by the herringbone markings on its hind femur. Another short-horned grasshopper familiar to anyone who has dissected a grasshopper in biology lab is the lubber grasshopper, *Brachystola* spp. This large grasshopper has short wings and is not capable of flight. Occasionally, these grasshoppers occur in great enough numbers to devastate large areas of cotton in West Texas.

Mantids

Order: Mantodea
(man-TOH-dee-uh)
mantis, soothsayer

Characteristics:
SIZE: 2 to 4 in. (5.0 to 10.0 cm)
SHAPE: Elongated
COLOR: Variable
ANTENNAE: Long and threadlike
MOUTHPARTS: Chewing
EYES: Compound
WINGS: Two pairs; lie straight down back
LEGS: **Forelegs modified for grasping;** second and third pair, walking; **spacing between front and second pair of legs much longer than between second and third pair**
MISCELLANEOUS: **Head triangular and movable**

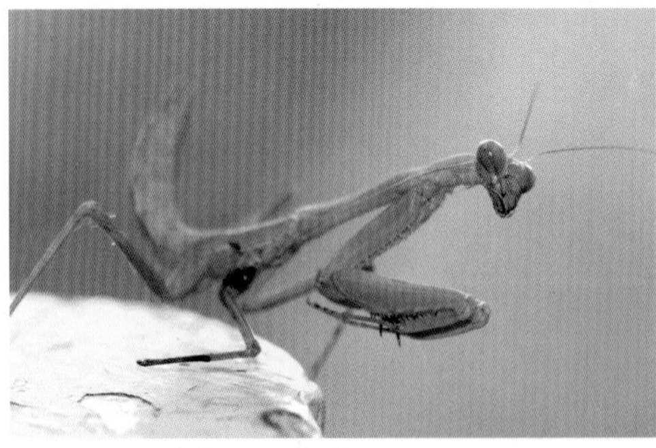

Mantid (by Joe Carter)

Mantid ootheca

Mantid head

HABITAT: Trees and shrubs
FOOD: Predaceous on insects and other small organisms
METAMORPHOSIS: Gradual (egg, nymph, adult)

Mantids, also known as "praying mantids" or "soothsayers," are large insects with powerful, distinctive grasping forelegs. Mantids also have a triangular head capable of turning to allow them to "look over their shoulder," which is rare in the insect world.

These stealthy insects lie in wait for unsuspecting insects to pass within reach. Male mantids are relatively uncommon because of the females' macabre habit of eating the male soon after (or sometimes during) copulation.

The female mantid lays about two hundred eggs in a mass called an ootheca that is attached to twigs, walls, and other structures. Mantids overwinter in the egg stage, and oothecae can be collected and placed in gardens or around the house for biocontrol of pest insects. If no insect prey is available, the first newly hatched mantids will eat the others as they emerge.

Cockroaches, Water Bugs

Order: Blattodea
(bluh-TOH-dee-uh)
blatta, an insect that shuns light

Cockroach, side view

Cockroach with ootheca (by Gary Brooks)

Characteristics:

SIZE: ½ to 3 in. (1.3 to 7.5 cm)
SHAPE: **Oval; flattened**
COLOR: **Typically various shades of brown**
ANTENNAE: Threadlike
MOUTHPARTS: Chewing
EYES: Compound
WINGS: Most winged; a few wingless
LEGS: Walking; **tibia possesses long, stiff bristles**
MISCELLANEOUS: **Head covered by pronotum and not visible from above; nocturnal**
HABITAT: **Dark, moist areas**
FOOD: Mostly decomposing materials
METAMORPHOSIS: Gradual (egg, nymph, adult)

Cockroaches in the house are probably one of the most disgusting sights homeowners witness. These insects are exquisitely designed for living with humans. Their flattened bodies allow them to slip into the smallest cracks of kitchen and bathroom cabinets, where they hide during daylight. They leave these protective hiding places under the cloak of darkness and are often seen only when someone turns on the lights.

Cockroach, top view

There are over three thousand species of cockroaches in the world. Most of these live outside and are important in recycling decomposing animal and plant materials. Five species, however, are specially adapted for life with humans. The oriental cockroach, *Blatta orientalis;* American cockroach, *Periplaneta americana;* brown-banded cockroach, *Supella longipalpa;* and smoky-brown cockroach, *P. fuliginosa,* can become pests under the right conditions.

The German cockroach, *Blattella germanica,* however, is dependent upon human activity and is the most common household pest in Texas. The female German cockroach produces an egg sac containing twenty to thirty eggs that she carries with her until just before the nymphs emerge. These insects breed continuously, and the entire life cycle is completed in about one hundred days.

Termites

Order: Isoptera
(y-SAHP-tur-uh)
iso, equal; *ptera,* wing

Characteristics:

SIZE: About $\frac{3}{8}$ in. (9.6 mm)
SHAPE: **Abdomen broadly joined at the thorax**
COLOR: **Typically cream colored**
ANTENNAE: **Beadlike and straight**
MOUTHPARTS: Chewing
EYES: Compound; present in all winged forms; present or absent in wingless forms
WINGS: **Absent in workers; present in reproductive forms that swarm; two pairs, equal size; fracture line at base**
LEGS: Walking
MISCELLANEOUS: Soft bodied
HABITAT: **Soil, wood; prefer high moisture content**
FOOD: Wood (cellulose); form a mutualistic symbiotic relationship with protozoans
METAMORPHOSIS: Gradual (egg, nymph, adult)

Termites are social insects that feed on wood and other cellulose materials. In the wild, their activities help break down and recycle dead trees and other plant materials. Unfortunately, termites cannot distinguish between dead trees and the lumber used to build houses. These insects cause millions of dollars in damage to homes in Texas.

Most termites require high humidity to survive and are only seen in the spring when new colonies are formed or after they have caused damage to a wooden structure. Shortly after a spring rain, swarms of reproductive females and males leave the old colony and mate. Then the females seek a location to establish a new colony.

Many ants also disperse after rains, so it is very important to be able to distinguish them from termites. Ants have elbowed antennae and a restricted waist; termites have beadlike antennae, and their abdomen is broadly joined to the thorax.

The subterranean termite, *Reticulitermes* spp., is the most common termite to infest homes. These insects require wood-to-soil contact, and typically termite prevention or control is aimed at placing a barrier between the two. The drywood termite, however, occurs in the humid coastal regions of Texas. It does not require contact with the soil and can live in sound wood aboveground.

Desert termites, *Gnathamitermes tubiformans,* are common in pastures. These termites do not feed inside the plant but rather on surface tissue of both herbaceous and woody plants.

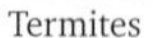
Termites

Winged termite

Desert termites

Earwigs

Order: Dermaptera
(dur-MAP-tur-uh)
derma, skin; *ptera,* wing

Characteristics:
SIZE: $\frac{1}{8}$ to $1\frac{1}{4}$ in. (3.2 mm to 3 cm)
SHAPE: **Cylindrical**
COLOR: Typically brown
ANTENNAE: Threadlike
MOUTHPARTS: Chewing
EYES: Compound
WINGS:
Forewings: **Very short; leathery**
Hind wings: Membranous; intricately folded under forewings
LEGS: Walking
MISCELLANEOUS: **Well-developed forceplike cerci; nocturnal**
HABITAT: **Under rocks and other structures during daytime**
FOOD: Scavengers; occasionally feed on plants; predaceous on other insects
METAMORPHOSIS: Gradual (egg, nymph, adult)

Earwig

Earwigs are readily recognized by their short forewings and forceplike cerci at the end of their abdomen. The inner edge of the cerci is straight in female earwigs but curved in males.

The common name "earwig" is derived from the superstition that these insects can crawl into a person's ear. This is not true. These insects are quite harmless, although they can inflict a slight pinch with their cerci. Earwigs do, however, possess a scent gland and can emit a rather foul-smelling fluid.

Earwigs prefer moist areas such as flower beds but may enter homes during hot, dry periods. They feed mainly on decomposing materials but occasionally on plants, and may cause considerable damage. They are nocturnal and seldom seen during the daytime.

A female earwig lays fifteen to eighty eggs in a nest, which she will guard for a short time even after the eggs have hatched. This prenatal care is unusual in the insect world. These insects overwinter as adults.

Earwig

Webspinners

Order: Embiidina
(ehm-bee-ih-DEE-nuh)
embios, lively

Characteristics:

SIZE: $\frac{1}{8}$ to $\frac{3}{8}$ in. (3.2 to 9.6 mm)
SHAPE: **Cylindrical**
COLOR: **Typically brown**
ANTENNAE: Threadlike
MOUTHPARTS: Chewing
EYES: Compound
WINGS: Females wingless; males winged or wingless
LEGS: Hind femora enlarged; **front tarsi swollen and modified for silk production**
MISCELLANEOUS: Cerci present
HABITAT: **Silken galleries**
FOOD: Scavenger
METAMORPHOSIS: Gradual (egg, nymph, adult)

Webspinners are most often noticed because of the silken webs they produce on the bark of trees, which can be mistaken as fungal growth. These small, cylindrical, brown insects are readily recognized by the swollen tarsi of their forelegs, which have glands that produce silk.

Webspinners live in colonies made up of a single female and her offspring. These colonies can be found in leaf litter, under stones, in cracks, in the soil, or under tree bark. Females are wingless and feed on decomposing materials; the males are winged and do not eat.

Both adults and nymphs are capable of producing silk, which serves to protect the gallery. When attacked, these small insects move backward between the narrow corridors on the tree trunk to elude the predator. There is usually one generation per year.

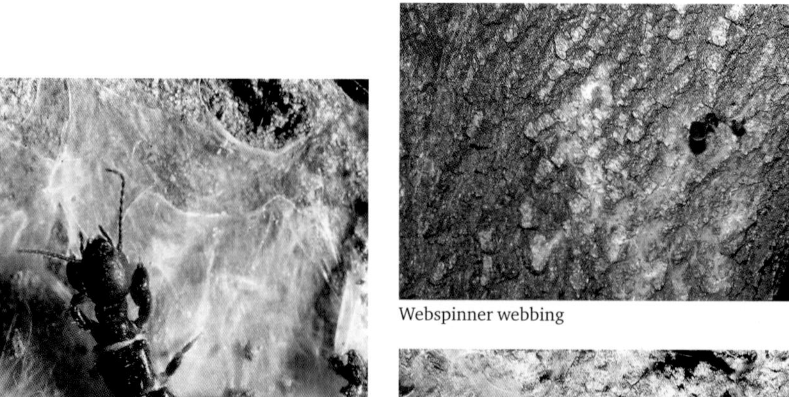

Webspinner webbing

Webspinner

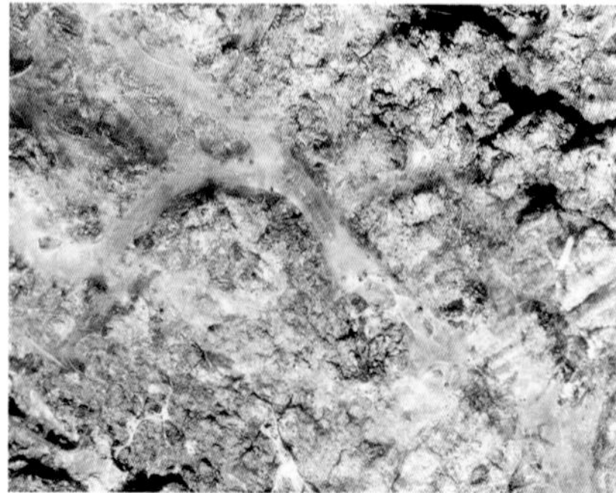

Webspinner in gallery

Stoneflies

Order: Plecoptera
(plee-KAHP-tur-uh)
pleco, folded; *ptera,* wing

Characteristics:
SIZE: ⅜ to 1⅝ in. (1.0 to 4.0 cm)
SHAPE: **Elongated; flattened**
COLOR: Yellowish brown
ANTENNAE: **Threadlike; at least half the length of the body**
MOUTHPARTS: Chewing; often nonfunctional
EYES: Compound
WINGS: **Soft; folded over abdomen**
LEGS: Walking
MISCELLANEOUS: **Two long, segmented tails**
HABITAT:
Adults: **Near water**
Nymphs: **Under stones in clean, well-oxygenated water**
FOOD:
Adults: Algae, lichens, and plants
Nymphs: Omnivorous; some plant feeders; some predaceous on small aquatic organisms
METAMORPHOSIS: Incomplete (egg, nymph, adult)

Stoneflies are found near clean, well-oxygenated water, such as streams and lakes. Adults are cylindrical insects that are somewhat flattened and possess two long, segmented tails. They are poor fliers and are often attracted to lights located near water.

The female stonefly carries her eggs on her abdomen before depositing them in water. The aquatic nymphs, also called naiads, typically live under stones, giving them their common name. The nymphs, which resemble adults but do not have wings, undergo up to twenty-two molts before reaching maturity.

These insects are very sensitive to water pollution and are often an early indicator of changes in the water chemistry. They are considered beneficial because some species are predaceous on other aquatic organisms and are also an important food source for fish.

Stonefly, top view

Stonefly, side view

Booklice, Barklice

Order: Psocoptera
(soh-KAHP-tur-uh)
psokos, gnawed; *ptera,* wing

Characteristics:

SIZE: $\frac{1}{32}$ to $\frac{3}{16}$ in. (0.8 to 4.8 mm)

SHAPE: Typically humpbacked; large head

COLOR: Booklice straw colored; barklice dark

ANTENNAE: Long and threadlike

MOUTHPARTS: Chewing

EYES: Compound; large and bulging

WINGS: Present or absent; when present, two pairs, **held rooflike; forewings larger than hind wings**

LEGS: Walking

HABITAT:

Booklice: **Indoors and outdoors; books, paper, furniture, stored grain, bird nests**

Barklice: **Outdoors; bark and leaves of trees**

FOOD:

Booklice: Scavengers; microorganisms

Barklice: Lichens; fungi

METAMORPHOSIS: Gradual (egg nymph, adult)

Booklice are tiny, wingless insects that are common in homes and warehouses, where they feed on stored grains, glue, book bindings, and other starchy materials. They are also found in bird nests feeding on detritus.

Barklice are much larger than booklice and are winged. They prefer moist environments and live in leaf litter, under rocks, and on tree bark. They are often found in clusters hidden under a thin layer of silk. They eat lichens and fungi growing on bark but do not damage trees.

Booklice

Barklice (by Gary Brooks)

Barklice

Chewing Lice

Order: Phthiraptera
(thehr-AHP-tur-uh)
Suborder: Mallophaga
(muh-LAH-fuh-guh)
mallo, wool; *phaga,* to eat

Chewing lice

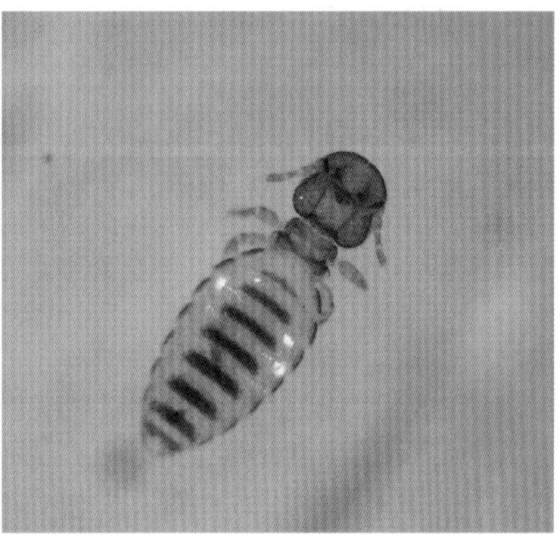
Chewing louse

Characteristics:
SIZE: **<⅛ in. (3.2 mm)**
SHAPE: **Head triangular, broader than thorax**
COLOR: Typically yellowish
ANTENNAE: Present; short; sometimes concealed
MOUTHPARTS: Chewing
EYES: Reduced or absent
WINGS: **Absent**
LEGS: Short, stout; one or two large claws
HABITAT: **Birds; mammals**
FOOD: Hair; feathers; skin
METAMORPHOSIS: Gradual (egg, nymph, adult)

Chewing, or biting, lice are small, wingless parasites of birds and mammals. They are distinguished from the sucking lice by their large triangular head that is wider than their thorax.

Chewing lice can be a problem on farm animals such as chickens, turkeys, cattle, and sheep. They spend their entire life on the animal feeding on hair, feathers, and dead skin. Their presence can cause irritation and result in an unthrifty animal. Chewing lice are spread among animals by contact.

Humans are not a preferred host of chewing lice, but livestock workers may become temporarily infested when handling infested animals. No known disease-causing organisms are transmitted by chewing lice.

Sucking Lice

Order: Phthiraptera
Suborder: Anoplura
(an-uh-PLUR-uh)
anaoplus, unarmed; *plura,* tail

Characteristics:
SIZE: **<$\frac{1}{8}$ in. (3.2 mm)**
SHAPE: **Head narrower than thorax**
COLOR: Tan
ANTENNAE: Present
MOUTHPARTS: Piercing/sucking
EYES: Compound; reduced or absent
WINGS: **Absent**
LEGS: Tarsi 1-segmented and fits into a thumblike structure used to hang on to host
HABITAT: **Mammals**
FOOD: **Blood**
METAMORPHOSIS: Gradual (egg, nymph, adult)

Sucking lice are small, parasitic insects of a wide range of mammals. They are distinguished from chewing lice by their pointed head that is narrower than their thorax.

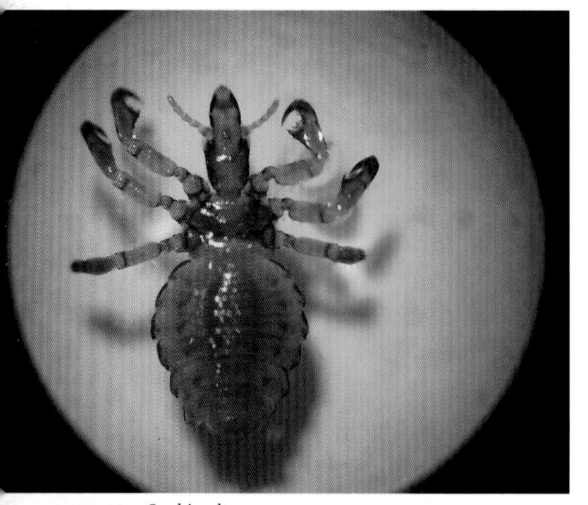

Sucking louse

Sucking lice feed on the blood of their host and can cause irritation to livestock. It is not uncommon to find cattle rubbing off large patches of hair trying to relieve the itching. Greatest numbers generally occur in the cooler months of the year.

Crab lice, *Pthirus pubis;* head lice, *Pediculus humanus capitis;* and body lice, *P. humanus humanus,* feed on humans. Being blood feeders, some species of sucking lice transmit diseases. It has been estimated that more than 3 million soldiers died in World War I as a result of body louse–borne typhus.

Today, head lice are a common problem among school-aged children. The lice are spread by contact, probably as a result of kids playing or borrowing each other's personal items. It is not known if head lice can transmit disease-causing organisms.

Sucking louse

True Bugs, Aphids, Cicadas, and Hoppers

Order: Hemiptera
(heh-MIHP-tur-uh)
hemi, half; *ptera*, wing

Characteristics:

SIZE: $\frac{1}{16}$ to 2 $\frac{1}{5}$ in. (0.2 to 6.4 cm)
ANTENNAE: Mostly threadlike
MOUTHPARTS: **Piercing/sucking**
EYES: Compound; generally present; ocelli variable
WINGS: Generally present; few wingless
LEGS: Variable: may be modified for walking, swimming, or grasping
HABITAT: Most terrestrial on plants; some aquatic
FOOD: Plant feeders; predaceous on small arthropods; few parasitic on other animals
METAMORPHOSIS: Gradual (egg, nymph, adult)

The hemipterans are a large, very diverse group that at one time was separated into two orders: Hemiptera, true bugs; and Homoptera, cicadas, hoppers, aphids, and others. Authorities now place both groups in the order Hemiptera with three suborders.

Although all insects are often referred to as "bugs," the correct use of the term is reserved for members of the suborder Heteroptera. True bugs can typically be identified by the distinctly two-textured forewings that lie flat over the back. They have piercing/sucking mouthparts that originate in the front of the head and are held flat under the body when not in use.

Members of the suborder Auchenorrhyncha include the cicadas and hoppers. These insects have piercing/sucking mouthparts that originate at the back of the head. Although some species are wingless, most have wings that are the same texture throughout and typically held rooflike.

Members of the suborder Sternorrhyncha include the aphids, scale insects, and whiteflies. These insects may be winged or wingless and have piercing/sucking mouthparts that originate between the front coxae. Many members of this group are important pests of field crops or ornamental plants.

Common families of Hemiptera, suborder Heteroptera, with their preferred habitats and feeding characteristics

Family	*Aquatic*	*Plants*	*Ground*	*Plant feeders*	*Predators*	*Parasites*	*Page*
Ambush bugs: Phymatidae		X			X		74
Assassin bugs: Reduviidae		X			X	X (rare)	75
Backswimmers: Notonectidae	X				X		76
Bed bugs, bird bugs: Cimicidae						X	77
Burrowing bugs: Cydnidae			X	X			78
Damsel bugs: Nabidae		X			X		79
Giant water bugs: Belostomatidae	X				X		80
Lace bugs: Tingidae		X		X			81
Leaf-footed bugs: Coreidae		X		X			82
Minute pirate bugs: Anthocoridae		X			X		83
Plant bugs, leaf bugs: Miridae		X		X	X (rare)		84
Seed bugs, milkweed bugs: Lygaeidae		X		X			85
Stilt bugs: Berytidae		X		X	X (few)		86
Stink bugs: Pentatomidae		X		X	X (few)		87
Toad bugs: Gelastocoridae	X				X		88
Water boatmen: Corixidae	X				X		89
Waterscorpions: Nepidae	X				X		90
Water striders: Gerridae	X				X		91

Common families of Hemiptera, suborders Auchenorrhyncha and Sternorrhyncha, with their preferred habitats

Family	*Grasses*	*Herbaceous plants*	*Trees and shrubs*	*Page*
Aphids, plantlice: Aphididae	X	X	X	92
Cicadas: Cicadidae			X	93
Froghoppers, spittlebugs: Cercopidae		X	X	94
Leafhoppers: Cicadellidae	X	X	X	95
Mealybugs: Pseudococcidae	X	X	X	96
Planthoppers: superfamily Fulgoroidea	X	X	X	97
Scale insects: superfamily Coccoidea	X	X	X	98
Treehoppers: Membracidae			X	99
Whiteflies: Aleyrodidae		X	X	100

Ambush Bugs

Family: Phymatidae
(fy-MAT-ih-dee)

Ambush bug (by Gary Brooks)

Characteristics:

SIZE: **$\frac{1}{4}$ to $\frac{1}{2}$ in. (6.4 to 12.8 mm)**
SHAPE: Flat; elongated; **abdomen widens toward the rear**
COLOR: **Usually yellowish brown**
ANTENNAE: Slightly clubbed; 4-segmented
MOUTHPARTS: Piercing/sucking; 3-segmented; arise from the lower front of the head
EYES: Compound; two ocelli
WINGS: Forewings two textured; **narrower than abdomen**
LEGS: **Front pair enlarged and modified for grasping**
MISCELLANEOUS: **Often resemble the flowers in which they hide**
HABITAT: Plants; **flowers**
FOOD: Predaceous on other arthropods

Ambush bugs hide within a cluster of flowers and are often not seen without careful observation. Their cryptic coloration and ability to remain motionless allow them to capture flies, bees, and other insects visiting the host plant for pollen and nectar.

Once they are located, these predaceous insects are easily recognized by the greatly enlarged forelegs and distinctive abdomen shape. Ambush bugs are not known to bite when handled.

Ambush bug with fly (by Barry Lambert)

Assassin Bugs

Family: Reduviidae
(reh-joo-VEE-ih-dee)

Characteristics:

SIZE: $\frac{1}{4}$ to ≥ $1\frac{1}{3}$ in. (0.6 to 3.4 cm)

Shape: **Variable; abdomen often widens in the middle and extends beyond the wings**

COLOR: Variable

ANTENNAE: Threadlike; 4- or 5-segmented

MOUTHPARTS: Piercing/sucking; **stout, short, and curved; fitting into a groove between forelegs;** 3-segmented; arise from the lower front of the head

EYES: Compound; ocelli, usually two, rarely absent

WINGS: Forewings two textured

LEGS: Walking; front femora generally thickened; tarsi 3-segmented

MISCELLANEOUS: **Head elongated, narrow, with a transverse groove between the eyes; a few species resemble walkingsticks**

HABITAT: Plants

FOOD: Most predaceous on other arthropods; some parasitic, feeding on blood

Assassin bugs are common insects in a variety of plant habitats. Although quite variable in size, color, and shape, these insects are readily recognized by the short, curved beak, which fits into a groove between the forelegs. The head is also generally slender; the eyes bulge and have a distinct groove between them.

Wheel bug, *Arilus cristatus*

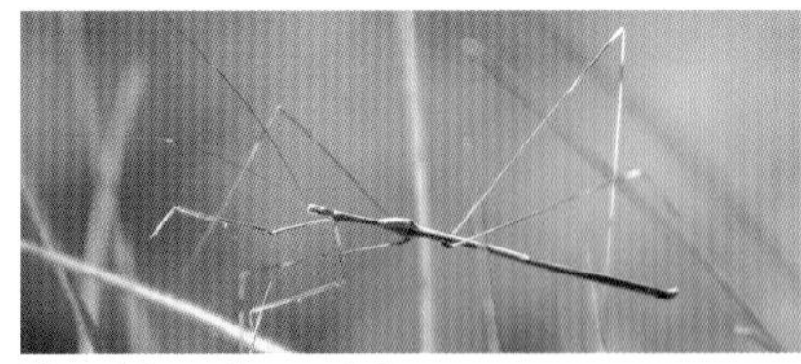

Thread-legged assassin bug

Assassin bugs emerging (by Gary Brooks)

Most species are predators of any insect they are able to catch. Many are adapted to blend with their surroundings. The thread-legged bugs of the subfamily Emesinae resemble small walkingsticks and often hide in grassy areas. The wheel bug, *Arilus cristatus*, is common in trees and can inflict a very painful bite.

Members of the genus *Triatoma* are known as "kissing bugs" for their habit of biting sleeping people on the lips. These parasitic insects occur in the southern United States and southward and are known to transmit the pathogen that causes Chagas disease.

Assassin bug mouthparts

Backswimmers

Family: Notonectidae
(noh-tuh-NEHK-tih-dee)

Characteristics:

SIZE: ~$\frac{1}{2}$ in. (12.8 mm)
SHAPE: **Boat shaped; back convexed**
COLOR: **Back, light colored; underside, lime green**
ANTENNAE: Short and inconspicuous; 4-segmented
MOUTHPARTS: Piercing/sucking; arise from the lower front of the head
EYES: Compound
WINGS: Forewings two textured
LEGS: **Forelegs modified for grasping; hind legs longer than other pairs and modified for swimming**
MISCELLANEOUS: **Swim on their back; generally seen with hind legs held outstretched like oars**
HABITAT: **Aquatic; lakes, ponds, and other calm water areas**
FOOD: Predaceous on other aquatic arthropods

Backswimmers resemble water boatmen but, as the name implies, swim upside down. They are well adapted for this lifestyle. Their back is shaped like the keel of a boat, allowing for quick movement through the water. Their back is also light colored, making it difficult for predators to see from underneath. Seen from above, the insect's underside is generally lime green and blends well with the surface of the water.

These fierce predators wait underwater for an unsuspecting insect to swim overhead. When at rest, the tip of the abdomen pierces the surface of the water and the head remains underwater at a sharp angle.

These insects can inflict a painful bite.

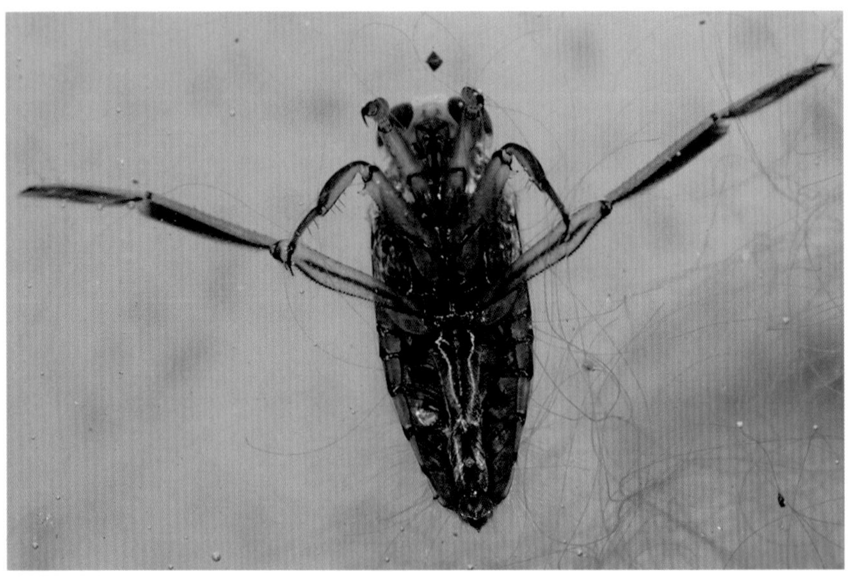

Backswimmer

Bed Bugs, Bird Bugs

Family: Cimicidae
(sih-MIH-suh-dee)

Characteristics:
SIZE: ~$\frac{1}{4}$ in. (6.4 mm)
SHAPE: **Flattened; oval**
COLOR: **Usually reddish brown**
ANTENNAE: Long; slender
MOUTHPARTS: Piercing/sucking; arise from the lower front of the head
EYES: Compound; ocelli absent
WINGS: **Wingless**
LEGS: Walking
HABITAT: Mammals and birds
FOOD: Parasitic (blood)

Bed bugs and bird bugs are well adapted for their parasitic lifestyle. They are wingless and flattened, which allows them to hide in the cracks and crevices of the host's nest. They generally feed only at night and hide during the daytime.

The immature nymphal stage molts five times as it develops and must feed on blood between each molt. These nymphs can live several months between blood meals.

Most species of Cimicidae feed on birds, but a few feed on bats. The bed bug, *Cimex lectularius,* is common in hotels, houses, and other living quarters and feeds on humans. During the daytime, bed bugs hide in secluded places like the seams and folds of bedding materials, window and door casings, picture frames, and baseboards. Fortunately, bed bugs have not been shown to vector any disease-causing organisms. Some people, however, may be sensitive to the bite and develop an allergic reaction.

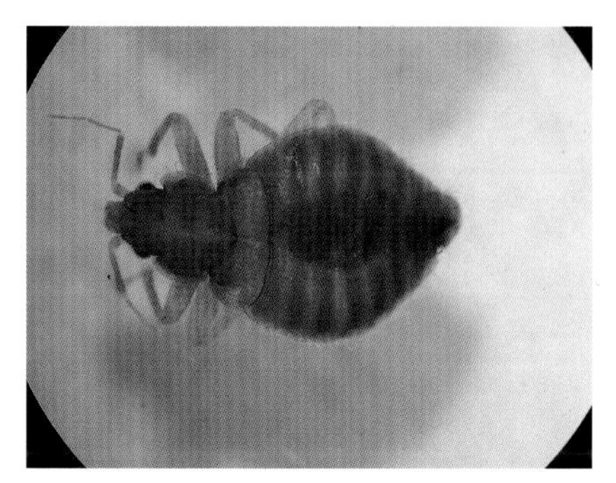

Bed bug, top view

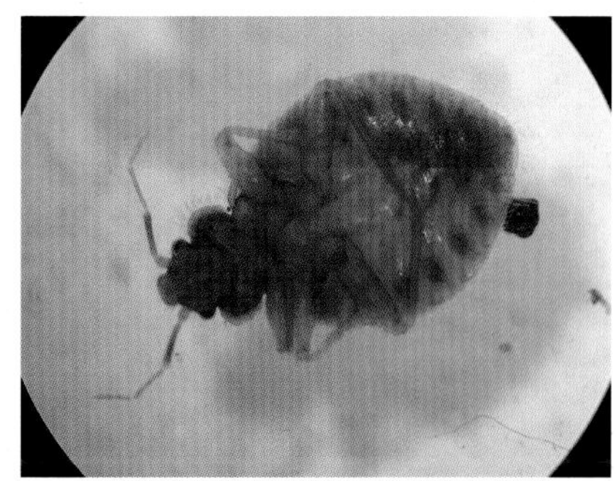

Bed bug, bottom view

Burrowing Bugs

Family: Cydnidae
(SIHD-nih-dee)

Characteristics:

SIZE: ¼ to ⅓ in. (6.4 to 8.5 mm)
SHAPE: **Oval**
COLOR: **Black**
ANTENNAE: Threadlike
MOUTHPARTS: Piercing/sucking; arise from the lower front of the head
EYES: Compound
WINGS: Forewings two textured
LEGS: **Forelegs often modified for digging; many spines on tibia**
MISCELLANEOUS: **Resemble stink bugs; large scutellum**
HABITAT: **Under rocks and around plant roots**
FOOD: Plant roots

Burrowing bugs are commonly found in sandy soils, under rocks, and around the bases of grasses and herbaceous plants. These oval insects resemble small stink bugs but can be separated from that family by the many spines located on the legs.

Burrowing bugs are attracted to artificial light and are frequent visitors to porch lights during the summer. The adult female burrowing bug lays her eggs in clumps in the soil and stays with the cluster of eggs until they hatch. Typically, there is only one generation per year, and they spend the winter as adults. Although they feed on plant roots, they seldom do enough damage to warrant control.

Burrowing bug

Damsel Bugs

Family: Nabidae
(NAB-ih-dee)

Damsel bug

Characteristics:

SIZE: ¼ to ⅜ in. (6.4 to 9.6 mm)
SHAPE: **Relatively slender and elongated**
COLOR: **Generally brown**
ANTENNAE: Threadlike; 4- or 5-segmented
MOUTHPARTS: Piercing/sucking; 4-segmented
EYES: Compound; ocelli present
WINGS: **Four long veins that end in several small cells around the margin**
LEGS: Walking; **front femur slightly enlarged;** tarsi 3-segmented
HABITAT: Generally on low vegetation
FOOD: Predaceous on other insects

Damsel bugs are typically light brown, slender insects with four long veins in their forewings that end in small cells on the margins of the wings.

Damsel bugs live in low-lying vegetation, where they feed on soft-bodied insects, such as small caterpillars, aphids, and various hoppers. Consequently, these insects are considered very beneficial in agricultural crops.

Damsel bug

Giant Water Bugs

Family: Belostomatidae
(behl-oh-stoh-MAT-ih-dee)

Characteristics:
SIZE: **1 to $2\frac{1}{2}$ in. (2.5 to 6.4 cm)**
SHAPE: **Oval; flattened**
COLOR: **Typically dark brown**
ANTENNAE: Short; inconspicuous; 4-segmented
MOUTHPARTS: **Short and stout;** arise from the lower front of the head
EYES: Compound
WINGS: Forewings two textured
LEGS: **Forelegs modified for grasping;** middle and hind legs flattened and modified for swimming
MISCELLANEOUS: **Resemble cockroaches; two short tails**
HABITAT: Aquatic; ponds and quiet pools in streams; **attracted to lights**
FOOD: Predaceous on other arthropods

Giant water bug

Male giant water bug with eggs

Giant water bug mouthparts

Some species of giant water bugs are among the largest insects in Texas, reaching up to two and one-half inches (6.4 cm) in length. They are oval and flattened and bear some resemblance to cockroaches. Their size and grasping forelegs easily distinguish these insects from most others.

Giant water bugs prefer calm water, where they prey on other insects, snails, and even small fish. The female of some species of the genus *Belostoma* lays her eggs on the back of a male, who will carry the eggs until they hatch.

These impressive insects are attracted to artificial lights and will "play dead" when disturbed. But beware: they can inflict a very painful bite if handled improperly.

Lace Bugs

Family: Tingidae
(TIHN-jih-dee)

Lace bug

Characteristics:
SIZE: $\frac{1}{16}$ to $\frac{1}{8}$ in. (1.6 to 3.2 mm)
SHAPE: Dorsally flattened; rectangular
COLOR: **Adults typically white; nymphs, brown or black and covered with spines**
ANTENNAE: Threadlike; 4-segmented
MOUTHPARTS: Piercing/sucking; arise from the lower front of the head
EYES: Compound
WINGS: **Lacelike**
LEGS: Walking; 1- or 2-segmented
HABITAT: Plants; **underside of leaves**
FOOD: Plant leaves

The name “lace bug” is very descriptive of these tiny bugs that are usually found in groups on the underside of leaves. Lace bugs are visible with the unaided eye, but to fully appreciate their beauty, observe them through a hand lens or magnifying glass. The adult body is covered by a lacy pronotum and wings, whereas the nymphs are much darker and very spiny.

Lace bugs are often pests on ornamental trees and shrubs, including sycamore, chrysanthemum, hawthorn, and azalea. Although they feed on the underside of leaves, the upper portion becomes blotched with yellow spots. They also produce honeydew and may be tended by ants that feed on the sugary substance.

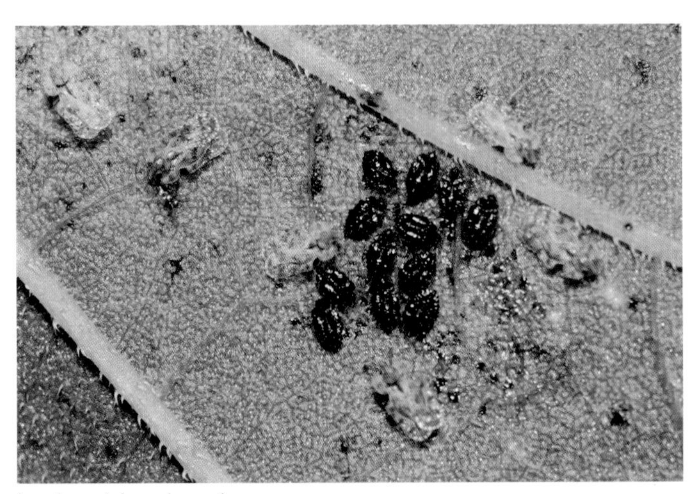
Lace bug adults and nymphs

Leaf-footed Bugs

Family: Coreidae
(kor-EE-ih-dee)

Characteristics:

SIZE: $\frac{1}{2}$ to 2 in. (1.3 to 5.1 cm)

SHAPE: **Elongated; head narrower and shorter than pronotum**

COLOR: Variable; generally brown

ANTENNAE: Threadlike; sometimes with dilated areas; 4-segmented

MOUTHPARTS: Piercing/sucking; 4-segmented; arise from the lower front of the head

EYES: Compound; two ocelli

WINGS: **Many parallel veins in forewings;** forewings two textured

LEGS: Walking; **hind tibia sometimes dilated and leaflike;** tarsi 3-segmented

HABITAT: Plants

FOOD: Mostly plant feeders; a few predaceous on other arthropods

Leaf-footed bug

Leaf-footed bug

Squash bug nymph, *Anasa tristis*

Squash bug adult, *Anasa tristis*

Leaf-footed bugs derive their common name from the leaflike swelling on the hind tibia of some species. Because this structure is not present on many species, a better identification is the numerous parallel veins in the membranous portion of the forewings. This characteristic is unique to this family.

Most coreids have a scent gland between the middle and hind legs and can secrete either a pleasant or a foul-smelling odor when handled. The squash bug, *Anasa tristis,* is a common garden pest that does not possess the dilated tibia.

Minute Pirate Bugs

Family: Anthocoridae
(an-thoh-KOR-ih-dee)

Characteristics:

SIZE: **$\frac{1}{16}$ to $\frac{1}{8}$ in. (1.6 to 4.8 mm)**
SHAPE: Oval
COLOR: **Mostly black and white**
ANTENNAE: Threadlike; 4-segmented
MOUTHPARTS: Piercing/sucking; 3-segmented; arise from the lower front of the head
EYES: Compound
WINGS: Forewings two textured; cuneus present; membrane with few or no closed cells
LEGS: Walking
HABITAT: **Flowers**
FOOD: Predaceous on small insects and insect eggs

Minute pirate bugs are very common insects found inside flower blooms. These tiny black-and-white insects are excellent predators that feed on arthropod eggs and small insects. The nymphs, usually orange in color, are found inside flowers. Both the adults and nymphs can also be found under tree bark and in grass.

The insidious flower bug, *Orius insidiosus,* is probably the most common species in this family and can inflict a somewhat painful bite for such a small insect.

Minute pirate bug

Plant Bugs, Leaf Bugs

Family: Miridae
(MEER-ih-dee)

Characteristics:
SIZE: $\frac{1}{4}$ to $\frac{3}{8}$ in. (6.4 to 9.6 mm)
SHAPE: Oval
COLOR: **Often brightly colored**
ANTENNAE: Threadlike; 4-segmented
MOUTHPARTS: Piercing/sucking; 4-segmented; arise from the lower front of the head
EYES: Compound; ocelli absent
WINGS: Forewings two textured; **cuneus present;** two closed cells; **dovetail posteriorly**
LEGS: Walking; tarsi 3-segmented
HABITAT: Plants
FOOD: Plant feeders; some predaceous on small arthropods

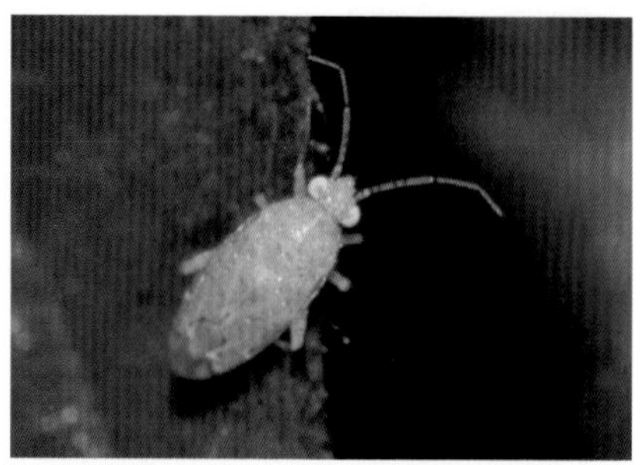
Cotton fleahopper, *Pseudatomoscelis seriatus*

Plant bug

Plant bug

Plant bug

The plant bug family is the largest in the suborder Heteroptera. Mirids can generally be recognized by looking at the insect from the side and noting the downward taper of the tip of the wings. Plant bugs are also one of the few families that have a cuneus. This structure can be seen as a triangular area on the outer edge of the wings between the hardened, opaque upper portion and the translucent lower portion.

Most plant bugs are plant feeders, but a few are predaceous on other small arthropods. The tarnish plant bug, *Lygus lineolaris,* can be a serious pest of legumes, flowers, and vegetables. The cotton fleahopper, *Pseudatomoscelis seriatus,* feeds on very small cotton flower buds. The damage may not be noticed until after the insect is gone and may seriously delay flowering.

Seed Bugs, Milkweed Bugs

Family: Lygaeidae
(ly-JEE-ih-dee)

Characteristics:

SIZE: $\frac{1}{4}$ to $\frac{5}{8}$ in. (6.0 to 16 mm)
SHAPE: **Variable**
COLOR: Variable; some black and white; often marked with red, white, or black spots or bands
ANTENNAE: Threadlike; located low on head; 4-segmented
MOUTHPARTS: Piercing/sucking; 4-segmented; arise from the lower front of the head
EYES: Compound; usually two ocelli
WINGS: Forewings two textured; **forewings with only four to five veins**
LEGS: Walking; front femur enlarged; tarsi 3-segmented
MISCELLANEOUS: **Head usually short and broad**
HABITAT: Plants, ground, and leaf litter
FOOD: Seed, plant sap; a few predaceous on other arthropods

Small milkweed bug, *Lygaeus kalmii*

Many members of the family Lygaeidae feed on seed. Some species, however, feed on milkweed plants, and a small number are predaceous on other insects and arthropods. The members of this large family are quite diverse in size, shape, and color. They may resemble mirids, coreids, or other members of families of true bugs but can be identified by the four to five unbranched veins in the membranous part of the forewings.

The chinch bug, *Blissus* spp., can be a serious pest of home lawns, small grains, corn, and other grasses. The big-eyed bug, *Geocoris* spp., is a common predator easily recognized by its bulging eyes. The red and black milkweed bug feeds on milkweed plants, which are toxic to most other animals.

Chinch bug, *Blissus* spp.

Big-eyed bug adult, *Geocoris* spp.

Big-eyed bug nymph, *Geocoris* spp.

Stilt Bugs

Family: Berytidae
(behr-IH-tih-dee)

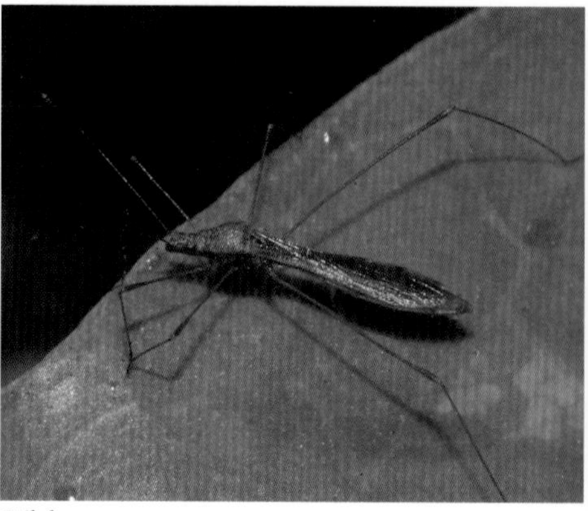
Stilt bug

Characteristics:

SIZE: ~$\frac{3}{8}$ in. (9.6 mm)
SHAPE: **Slender and elongated**
COLOR: **Brown**
ANTENNAE: Threadlike; **long, slender, and knobbed;** 4-segmented
MOUTHPARTS: Piercing/sucking; 4-segmented; arise from the lower front of the head
EYES: Compound; ocelli usually present
WINGS: Forewings two textured
LEGS: **Long and slender;** tarsi 3-segmented
HABITAT: Plants
FOOD: Generally plant feeders; some species partly predaceous on other insects

Stilt bugs are small, long-legged insects that are common on flowers, particularly wildflowers and weeds. Their antennae are also very slender and are as long as their legs. Stilt bugs might be confused with the thread-legged assassin bug. Both insects live on plants, but the thread-legged assassin bug is predaceous and can be recognized by its grasping forelegs.

Stilt bugs feed mainly on plant sap but may require protein from other insect eggs to mature properly.

Stilt bug

Stink Bugs

Family: Pentatomidae
(pehn-tuh-TAH-mih-dee)

Green stink bug

Stink bug eggs (by Gary Brooks)

Characteristics:

SIZE: $\frac{1}{3}$ to $\frac{1}{2}$ in. (8.5 to 12.8 mm)
SHAPE: **Shield shaped**; occasionally with spines located on the pronotum
COLOR: Typically green or brown; sometimes marked with red or yellow
ANTENNAE: Threadlike; 5-segmented
MOUTHPARTS: Piercing/sucking; 4-segmented; arise from the lower front of the head
EYES: Compound; two ocelli
WINGS: Forewings two textured
LEGS: Walking; tarsi 3-segmented
MISCELLANEOUS: **Large triangular area (scutellum) between wings**
HABITAT: Plants
FOOD: Most plant feeders; a few predaceous on other insects

Stink bug

Stink bug

Stink bugs are a large and common group of true bugs. They are generally easily recognized by their oval shape and the large triangular area between the wings known as the scutellum. Stink bugs can be quite variable in color; some are solid green or brown, others are very colorful, and others mimic tree bark. As the name suggests, many species produce a pungent odor when disturbed.

Most stink bugs are plant feeders, and several species are economically important pests. The harlequin stink bug, *Murgantia histrionica,* and the southern green stink bug, *Nezara viridula,* are common species in Texas and can cause plant damage.

Other species are predaceous and can be recognized by the thickened beak. Females lay barrel-shaped eggs in clusters on various plant parts.

Toad Bugs

Family: Gelastocoridae
(jehl-as-toh-KOR-ih-dee)

Characteristics:

SIZE: ~$\frac{1}{3}$ in. (8.5 mm)

SHAPE: **Toad shaped; short; oval; flattened**

COLOR: **Brown; sometimes camouflage**

ANTENNAE: Very short

MOUTHPARTS: Piercing/sucking; arise from the lower front of the head

EYES: **Compound, large, and projecting forward**

WINGS: Forewings two textured

LEGS: Walking

MISCELLANEOUS: **Resemble toads; able to hop**

HABITAT: **Semiaquatic**

FOOD: Predaceous on other arthropods

Toad bugs are remarkably similar to and easily mistaken for small toads. These humorous-looking insects live near or in shallow water, where they lie in wait for unsuspecting insect prey. Once the hapless insect is within reach, the toad bug jumps and seizes its victim.

Some species burrow into mud near the edge of water; others will hide in the sediments of the shallow edges. Many species can change colors to mimic their surroundings.

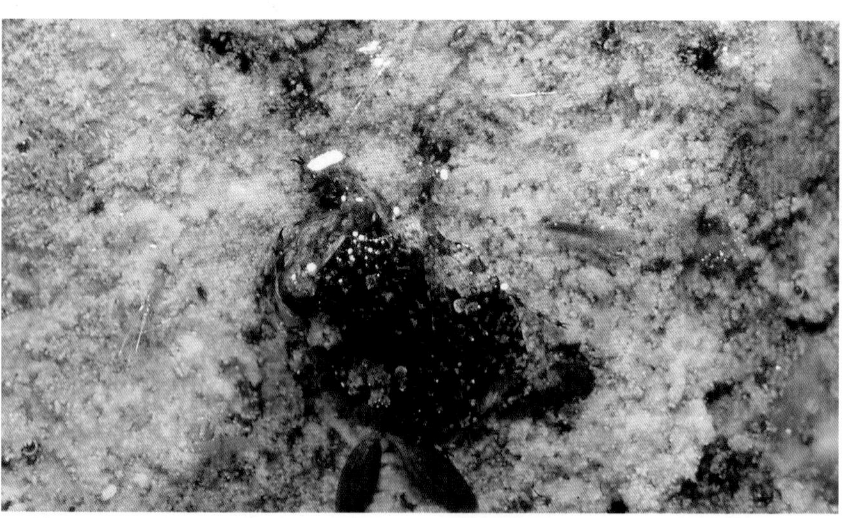

Toad bug

Toad bug

Toad bug

Water Boatmen

Family: Corixidae
(kor-IHKS-ih-dee)

Characteristics:
SIZE: ~$\frac{1}{2}$ in. (12.8 mm)
SHAPE: **Elongated; flattened on top**
COLOR: **Dark, mottled brown**
ANTENNAE: Small
MOUTHPARTS: Piercing/sucking; arise from the lower front of the head
EYES: Compound
WINGS: Forewings with same texture throughout; **many narrow, dark-colored bands**
LEGS: **Front tarsi scooplike for gathering food; hind legs elongated, function as oars**
HABITAT: **Fresh, still water**
FOOD: Algae; predaceous on other insects

Water boatmen are common insects in calm, fresh water. These insects can be recognized by the mottled brown pattern on their wings and their oarlike back legs. The water boatmen's abdomen is concave under the wings. Air stored in this cavity extends the time water boatmen can spend underwater.

Water boatmen dive to the bottom, where they feed on detritus, insects, and other small aquatic organisms. Unlike most aquatic true bugs, water boatmen do not bite.

Water boatmen are attracted to artificial light and are common visitors to porch lights during the summer months.

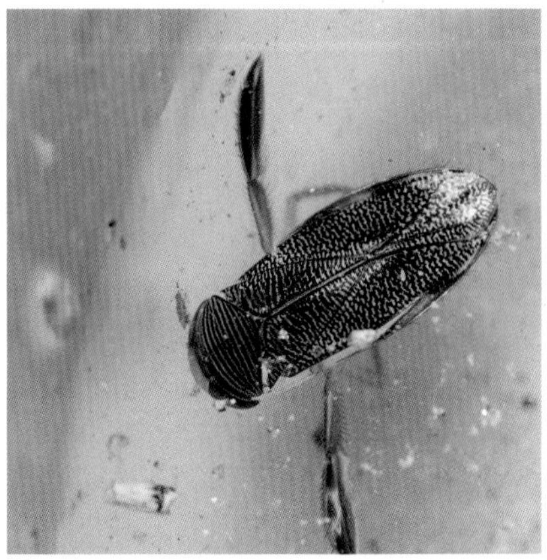

Water boatman

Water boatman

Waterscorpions

Family: Nepidae
(NEH-pih-dee)

Characteristics:

SIZE: $\frac{3}{4}$ to 3 in. (1.9 to 7.6 cm)
SHAPE: **Elongated; cylindrical; flattened**
COLOR: Tan
ANTENNAE: Small; 3-segmented
MOUTHPARTS: Piercing/sucking; arise from the lower front of the head
EYES: Compound
WINGS: Forewings two textured
LEGS: **Forelegs modified for grasping; hind legs adapted for swimming**
MISCELLANEOUS: **Long breathing tube at tip of abdomen**
HABITAT: **Aquatic**
FOOD: Predaceous on other insects and small aquatic organisms

Waterscorpions have two long breathing tubes at the end of the abdomen that allow the insects to stay submerged for extended periods of time. These structures are not retractable and give waterscorpions a somewhat scorpion-like appearance. Most species are long and slender, but a few are oval and resemble giant water bugs but can be identified by the much longer breathing tubes.

Waterscorpions' grasping forelegs are used to catch and restrain insect prey. These insects can inflict a painful bite if handled.

Waterscorpion (by Curtis Williams)

Water Striders

Family: Gerridae
(JEHR-ih-dee)

Characteristics:

SIZE: ~1 in. (2.5 cm)

SHAPE: **Elongated and cylindrical**

COLOR: Top, dark; underside, silvery white

ANTENNAE: Threadlike; 4-segmented

MOUTHPARTS: Piercing/sucking; arise from the lower front of the head

EYES: Compound

WINGS: **Reduced**

LEGS: **Forelegs modified for grasping; middle and hind legs long and adapted for swimming; middle legs close to hind legs;** tarsi 2-segmented

HABITAT: **Aquatic; calm, fresh water**

FOOD: Predaceous on other aquatic animals

Water striders are aquatic insects easily recognized by their grasping short forelegs and long middle and hind legs. The tarsi of the rear legs have fine hairs, which increase the area of contact with the surface of the water, allowing the insect to "skate" on the water surface.

Water striders are typically found in groups on calm water. These insects are predaceous on insects and other small aquatic organisms.

Water striders

Aphids, Plant Lice

Superfamily: Aphididae
(ay-FIHD-ih-dee)

Aphid giving birth

Characteristics:

SIZE: $\frac{1}{16}$ to $\frac{1}{4}$ in. (1.6 to 6.4 mm)
SHAPE: Pear shaped
COLOR: Variable
ANTENNAE: Threadlike and relatively long; 6-segmented
MOUTHPARTS: Piercing/sucking
EYES: Compound
WINGS: **Present or absent;** when present, transparent; held rooflike
LEGS: Walking
MISCELLANEOUS: Soft bodied; **most species possess a pair of cornicles near the end of the abdomen; some covered with waxy powder**
HABITAT: Plants: leaves, stems, flowers, some roots
FOOD: Plant feeders

Aphids are probably the most common insect pests on ornamental and crop plants. They feed on the plant, and many species are also vectors of plant disease-causing organisms. Most aphid species can be easily recognized by the presence of exhaust pipe–like structures, called cornicles, located at the rear of the abdomen. Defensive chemicals may be released from these structures when the insect is disturbed.

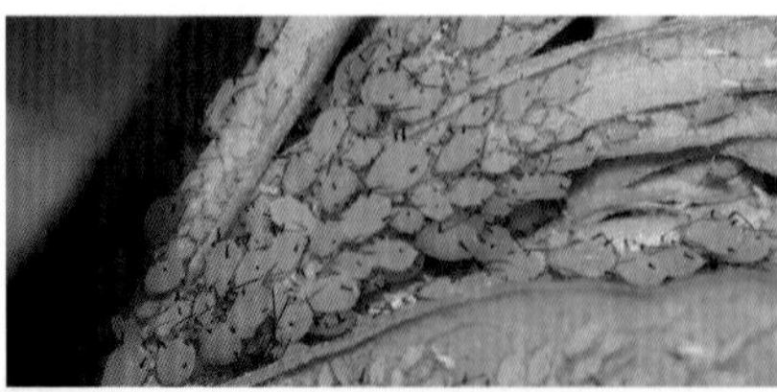
Aphids

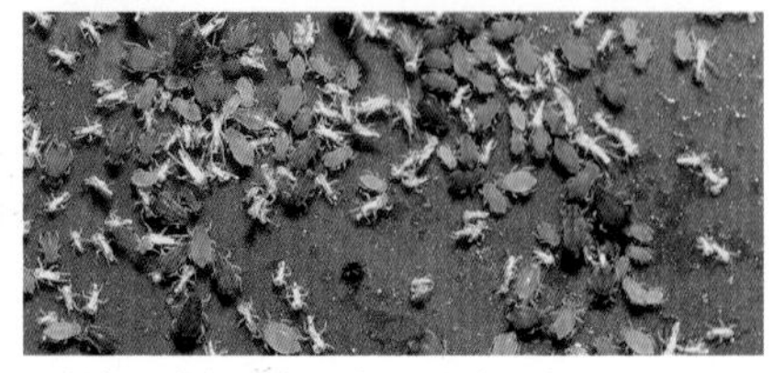
Aphids and shed skins (by Gary Brooks)

Aphids

Aphids have a rather complex life cycle. Several generations per year are common, and during the spring and summer females give birth, without mating, to live females rather than lay eggs. At other times, typically in the fall, males are produced, mating occurs, and eggs are produced.

Honeydew is a clear, sweet liquid secreted from the anus of aphids. Ants are commonly found feeding on the nutrient-rich excretions in aphid colonies. These ants benefit from the aphids and, in return, defend the aphids from predators, such as ladybird beetle larvae.

Honeydew is also an excellent food source for sooty mold. Leaves, car hoods, and any other place where the honeydew lands will often turn black with mold.

Cicadas

Family: Cicadidae
(sih-KAY-dih-dee)

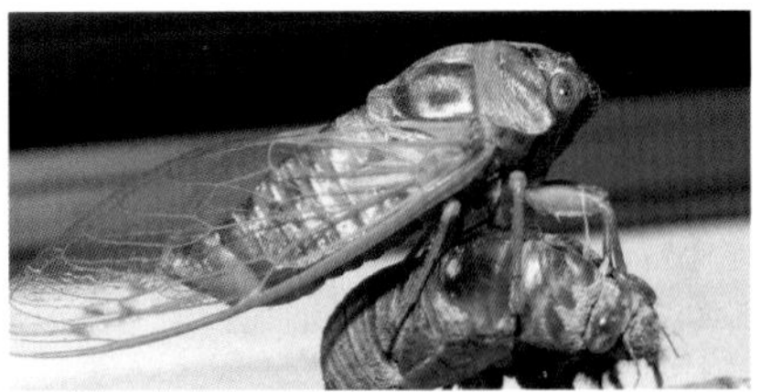
Cicada adult

Characteristics:
SIZE: **1.2 to 2 in. (1.3 to 5.1 cm)**
SHAPE: **Head wide and blunt**
COLOR: Variable; generally lime green or brown
ANTENNAE: **Short and bristlelike;** arise in front of eyes
MOUTHPARTS: Piercing/sucking
EYES: **Compound, prominent, and bulging;** three ocelli, forming a triangle
WINGS: Transparent; veins not reaching wing margins; held rooflike
LEGS: Walking; tarsi 3-segmented
MISCELLANEOUS: Males possess sound-producing organ
HABITAT:
Adults: Trees and shrubs
Nymphs: Underground
FOOD:
Adults: Plant sap
Nymphs: Root sap

Cicadas are often incorrectly called locusts, which are actually migratory grasshoppers. These common summertime visitors are easily recognized by their physically large size and blunt heads. Adults and nymphs feed on plant sap but typically do not cause serious damage. The female, however, will slash small twigs with her ovipositor to lay eggs, making them susceptible to breaking during high winds.

After the nymph emerges, it falls to the ground and burrows into the soil. This stage will last from one to nearly seventeen years depending on the species. After development is complete, the nymph emerges from the ground and climbs a tree, where the adult emerges through a slit in the nymphal skin. The old, discarded skin is a common reminder of their presence.

There are six species of periodical cicadas of the genus *Magicicada.* Three species have a thirteen-year life cycle, and three species have a seventeen-year life cycle. In Texas, the thirteen-year cicadas emerge between April and July. The more common dog-day cicadas emerge during the hot days of July and August after spending two to five years underground as nymphs.

Only male cicadas make sound by way of special abdominal muscles that vibrate rapidly against a drumlike structure on the underside of the abdomen. They "sing" only during the daytime and most often near sunset.

Cicada nymph

Cicada adult emerging

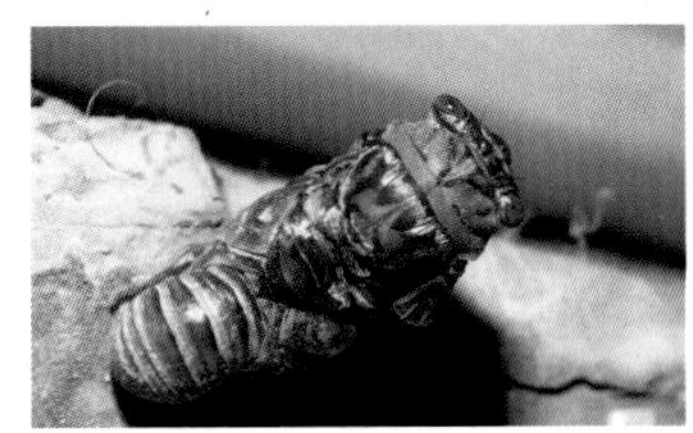
Cicada adult emerging

Froghoppers, Spittlebugs

Family: Cercopidae
(sur-KAH-pih-dee)

Characteristics:

SIZE: $\frac{1}{8}$ to $\frac{3}{8}$ in. (3.2 to 9.5 mm)

SHAPE: **Somewhat resemble frogs; oval widening toward the rear; head flat on top and usually narrower than pronotum**

COLOR: Typically brownish

ANTENNAE: Bristlelike; arise in front of and between eyes

MOUTHPARTS: Piercing/sucking

EYES: Compound; two ocelli

WINGS: Present; held rooflike

LEGS: Walking; **apex of tibia has circlets of spines;** hind tibia with one or two stout spines; 3-segmented

HABITAT: Plants

Froghopper, side view

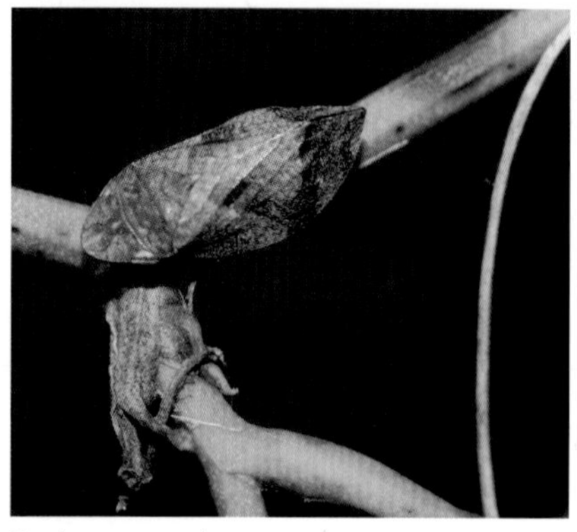

Froghopper, top view

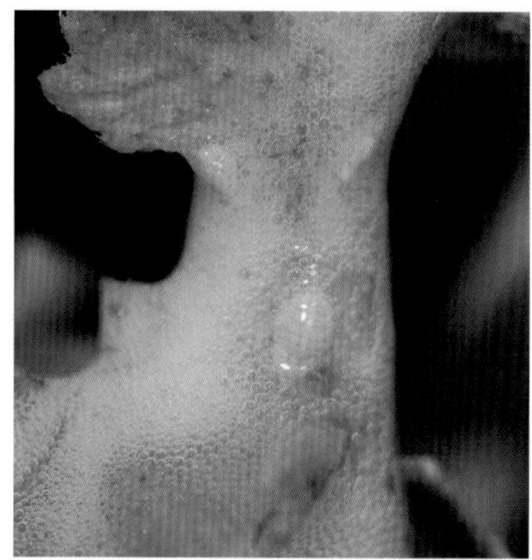

Spittlebug spittle

Spittlebug nymphs

FOOD: Grass and herbaceous plants; a few species attack trees

Froghoppers are common insects found in grassy areas and on herbaceous plants. Adults are typically brown, will jump when disturbed, and somewhat resemble small frogs. Nymphs are more commonly noticed because they produce and surround themselves with froth produced by forcing air through a thick, sugary solution released through the anus. This "spittle" is common on grasses, wildflowers, and other herbaceous plants in the spring.

Froghoppers and their nymphs, spittlebugs, generally do not cause plant damage, but some species are considered pests of clovers.

Leafhoppers

Family: Cicadellidae
(sih-kuh-DEHL-ih-dee)

Characteristics:

SIZE: $\frac{1}{16}$ to $\frac{1}{2}$ in. (1.6 to 12.8 mm)
SHAPE: **Elongated; body usually tapers posteriorly**
COLOR: **Many with colorful patterns on forewings**
ANTENNAE: Short and bristlelike; arise in front and between the eyes
MOUTHPARTS: Piercing/sucking
EYES: Compound; two ocelli
WINGS: Present; held rooflike
LEGS: Walking; **hind tibia with two parallel rows of spines;** 3-segmented
HABITAT: **All types of plants**
FOOD: Leaves and stems; almost all types of plants

The leafhoppers are the largest group in the suborder Auchenorrhyncha. They are typically cylindrical, are often very colorful, and tend to walk sideways when disturbed. The two parallel rows of spines are a definitive characteristic but are difficult to see without the aid of a hand lens or microscope.

Leafhoppers are among the most destructive groups of insects because, as a group, they feed on almost all types of plants. Damage can occur from the leafhopper's removing sap, plugging the plant's vascular system during feeding, and laying eggs in small twigs. This group is also known to transmit many plant disease-causing organisms.

Leafhopper

Mealybugs

Family: Pseudococcidae
(soo-doh-KAHKS-ih-dee)

Characteristics:

SIZE: $\frac{1}{16}$ to $\frac{1}{5}$ in. (1.6 to 5.1 mm)
SHAPE: **Oval and flattened**
COLOR: **Grayish or white**
ANTENNAE: Present; short; 1- to 9-segmented
MOUTHPARTS: Piercing/sucking
EYES: Compound
WINGS: **Females wingless; males winged**
LEGS: Walking
MISCELLANEOUS: **Covered in a white waxy substance; often with tail-like filaments**
HABITAT: Plants
FOOD: Plants

If you have houseplants, then you are probably familiar with mealybugs. These small fuzzy insects are pests of many plant species and can do considerable harm. They are closely related to the scale insects. The wingless females have legs and are covered in a powdery wax substance, which gives them the appearance of being "mealy." Males are winged but do not have functional mouthparts and die shortly after mating.

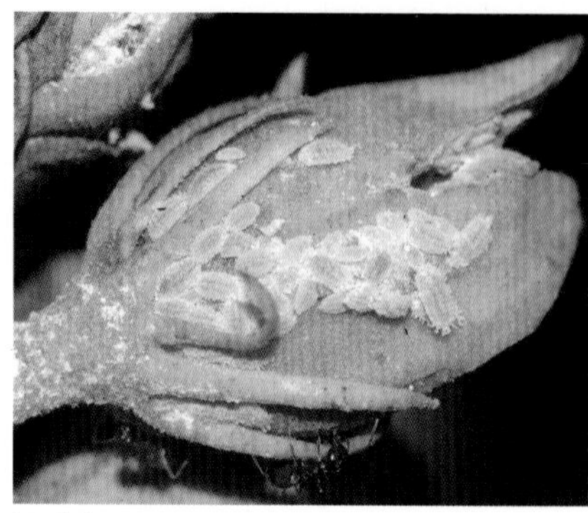

Mealybugs

Mealybugs may cause damage by feeding on host plant tissues and may even kill the plant if left unchecked. Mealybugs also inject toxins or plant pathogens that can cause premature leaf drop and dieback.

Mealybugs produce large quantities of honeydew that can create a shiny, sticky film on plant leaves and serve as a breeding area for sooty mold. Some historians believe that the manna from heaven that God provided for the Israelites was actually honeydew from the mealybug, *Trabutina mannipara.*

Mealybugs

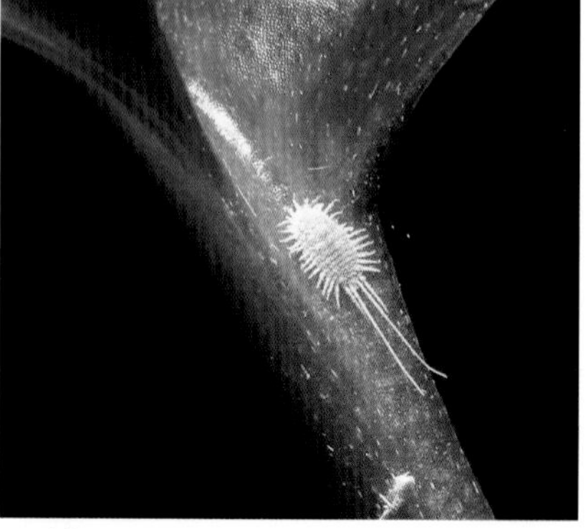

Mealybugs on ivy

Planthoppers

Superfamily: Fulgoroidea
(fuhl-guh-ROY-dee-uh)

Characteristics:
SIZE: Very small to ½ in. (12.5 mm)
SHAPE: Variable
COLOR: Variable
ANTENNAE: **Short and bristlelike; arise from sides of head beneath the eyes**
MOUTHPARTS: Piercing/sucking
EYES: Compound
WINGS: Present; held rooflike
LEGS: Walking; tarsi 3-segmented
MISCELLANEOUS: **A few species with prolonged head**
HABITAT: Plants
FOOD: Plants

Acanaloniid planthopper

Cixiid planthopper

Planthoppers are actually several families that constitute the superfamily Fulgoroidea. The more common members are the cixiid planthoppers (Cixiidae), fulgorid planthoppers (Fulgoridae), flatid planthoppers (Flatidae), and acanaloniid planthoppers (Acanaloniidae). Although diverse, members of this superfamily have antennae that arise from the sides of the head and beneath the eyes, and their hind tibia have only a few spines. These features distinguish them from other Auchenorrhyncha families.

Flatid planthopper

Most planthoppers feed on the phloem tissue of plants, and a few species can cause serious damage. Like many other Auchenorrhyncha, many species produce honeydew.

Scale Insects

Superfamily: Coccoidea
(kahks-OY-dee-uh)

Characteristics:

SIZE: $\frac{1}{16}$ to $\frac{1}{4}$ in. (1.6 to 12.8 mm)
SHAPE: Variable
COLOR: Variable
ANTENNAE: Often lacking in females; present in males
MOUTHPARTS: Piercing/sucking; lacking in males
EYES: Compound
WINGS: **Females wingless; males have one pair**
LEGS: **Females legless;** males and first instar nymphs with legs
MISCELLANEOUS: **Males resemble small gnats; female's and nymph's body covered by a waxy substance that forms a scale**
HABITAT: Plants
FOOD: Plants

The scale insects are one of the more destructive insect groups. This superfamily is composed of several families that are similar in physical characteristics,

Euonymus scale

Euonymus scale damage

Lecanium scale on live oak

biology, and ecology. The first instar nymphs, called "crawlers," have legs and antennae. This mobile stage seeks a new place to feed, attaches to the plant, sheds its legs, and secretes a waxy coating over its body. The adult females are wingless, legless, and cannot crawl.

Male scales resemble tiny gnats and have only one pair of wings. They do not have mouthparts, do not feed, and die shortly after mating. Males are rarely observed.

Soft scales (family Coccidae) have a larger side profile, and the waxy shell cannot be separated from the insect. The armored scales (family Diaspididae) have a lower side profile, and the covering can be removed. Both families are common pests of many ornamental plants and trees.

Cochineal insects, *Dactylopius coccus* (family Dactylopiidae), live on prickly pear cacti and are thought to have been a source of purple dye for Native Americans.

Cochineal scale on prickly pear cactus

Treehoppers

Family: Membracidae
(mehm-BRAS-ih-dee)

Treehopper

Characteristics:

SIZE: $\frac{1}{4}$ to $\frac{1}{3}$ in. (6.4 to 8.5 mm)

SHAPE: **Pronotum prolonged over abdomen; often humpbacked or thorn shaped**

COLOR: Greenish or brown

ANTENNAE: Short and bristlelike; arise in front of and between eyes

MOUTHPARTS: Piercing/sucking

EYES: Compound; two ocelli

WINGS: Present; often hidden by the enlarged pronotum; held rooflike

LEGS: Walking; tarsi 3-segmented

HABITAT: Plants

FOOD: Mostly trees and shrubs; some nymphs feed on herbaceous plants

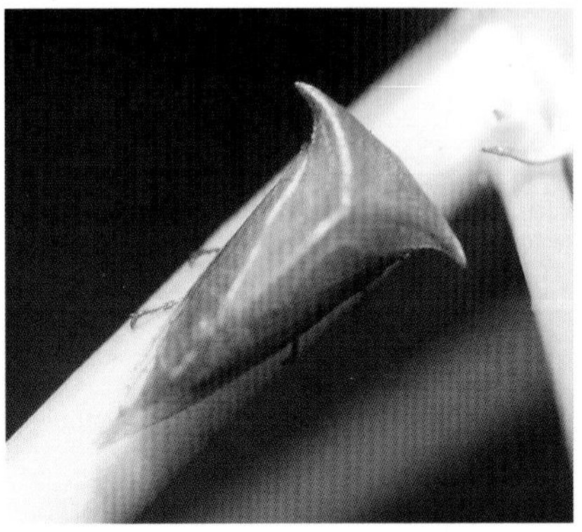
Treehopper

Treehoppers are recognizable by their elongated pronotum, which extends backward over the abdomen. Many species have bizarre shapes and often resemble thorns, allowing them to blend in with the plant. This form of mimicry provides protection against predators.

Treehopper nymphs often congregate in small groups that are tended by ants seeking the sugar-rich honeydew. Most species do not cause appreciable plant damage. The nymphs of the buffalo treehopper, *Stictocephala bizonia,* however, feed on alfalfa and can cause economic damage.

Treehopper

Treehopper (by James Lasswell)

Whiteflies

Family: Aleyrodidae
(al-ih-RAH-dih-dee)

Characteristics:
SIZE: **< $\frac{1}{16}$ in. (1.6 mm)**
SHAPE: **Resemble small moths**
COLOR: **Whitish**
ANTENNAE: Threadlike; 7-segmented
MOUTHPARTS: Piercing/sucking
EYES: Compound; two ocelli
WINGS: **Covered with whitish powder; held rooflike**
LEGS: Walking; tarsi 2-segmented
MISCELLANEOUS: **When disturbed, appear as floating ash**
HABITAT: Plants
FOOD: Plants

Whitefly nymphs

Whiteflies can be serious pests of greenhouse crops, citrus trees, many ornamental shrubs, and other plants. Upon close observation, adult whiteflies resemble small moths with a light, powdery, and waxy substance on their wings. The first instar nymphs are active, but subsequent instars are legless and immobile and may be mistaken for scales.

A female whitefly often will not stop feeding to lay eggs. Instead, she will simultaneously keep her beak inserted in the plant and lay her eggs in a circular pattern.

Plant damage occurs as the result of both sap removal from leaves and the production of honeydew. The honeydew is an excellent medium for sooty mold development, which causes the leaves to turn black.

Whitefly adults

Female whitefly laying eggs

Thrips

Order: Thysanoptera
(thy-suh-NAHP-tur-uh)
thysano, fringed; *plura,* tail

Characteristics:
SIZE: **$\frac{1}{32}$ to $\frac{1}{8}$ in. (0.8 to 3.2 mm)**
SHAPE: **Cigar shaped**
COLOR: Variable; **often straw colored; some with spotted wings**
ANTENNAE: Short
MOUTHPARTS: Rasping/sucking
EYES: Compound; ocelli, usually three
WINGS: **Narrow; fringed with hair**
LEGS: Walking
HABITAT: **Flowers; plant leaves**
FOOD: Plant feeders; predaceous on other small arthropods; some feed on fungal spores
METAMORPHOSIS: Gradual (egg, nymph, preadult, adult)

Thrips are very small, straw-colored, and cigar-shaped insects commonly found in flowers or near the veins of leaves. Upon close observation, the wings are very narrow and fringed with long, fine hairs. Thrips typically crawl with their abdomen pointed upward.

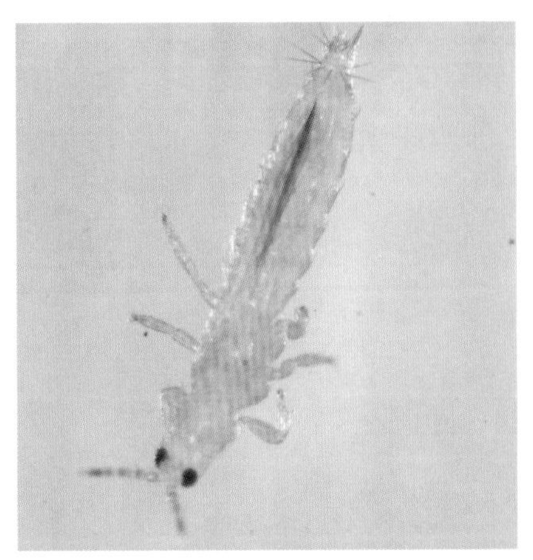

Thrips

Thrips, fringed wings

Thrips have an unusual life cycle. They may reproduce sexually, or the females may give birth to unfertilized eggs through a process known as parthenogenesis. Immature thrips go through four instars. The first two instars are active but do not have wing pads. Wing pads develop in the active third instar. The last instar, sometimes referred to as a preadult, does not feed and remains inactive until the adult emerges.

Plant-feeding thrips species can cause severe plant damage. They cut open leaf cells with their rasping mouthparts and suck the liquid. The empty cell causes the damaged leaf to curl and develop a silvery sheen.

The western flower thrips, *Frankliniella occidentalis,* and onion thrips, *Thrips tabaci,* are common pests of a wide variety of vegetables and other plants. The six-spotted thrips, *Scolothrips sexmaculatus,* however, is predaceous on plant-feeding mites.

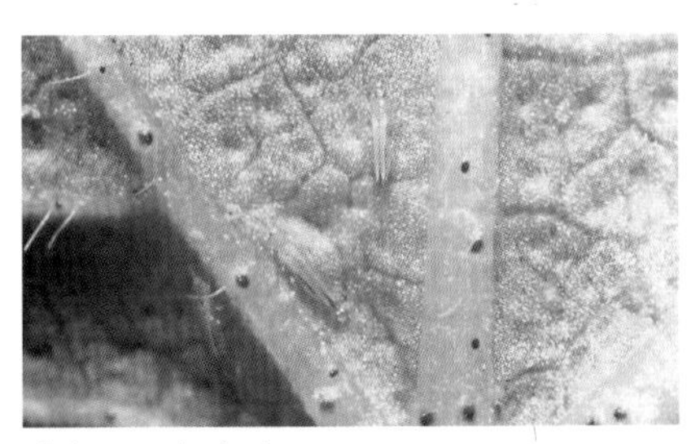

Thrips near leaf veins

Antlions, Dobsonflies, Lacewings

Order: Neuroptera
(nyoor-AHP-tur-uh)
neuron, sinew; *ptera,* wing

Characteristics:
SIZE: $\frac{1}{4}$ to 4 in. (6.4 to 100.0 mm)
ANTENNAE: Threadlike
MOUTHPARTS: Chewing
EYES: Compound
WINGS: **Translucent; many veins and cross veins**
LEGS: Mostly for walking; few for grasping
HABITAT:
Adults: Terrestrial
Larvae: Aquatic
FOOD: Most predaceous on other arthropods
METAMORPHOSIS: Complete (egg, larva, pupa, adult)

COMMON FAMILIES OF THE ORDER NEUROPTERA AND THEIR PREFERRED HABITATS

Family	*Aquatic*	*Ground*	*Plants*	*Page*
Antlions: Myrmeleontidae		L	A	103
Brown lacewings: Hemerobiidae			L, A	104
Dobsonflies (hellgrammites): Corydalidae	L		A	105
Green lacewings: Chrysopidae			L, A	106
Mantidflies: Mantispidae			L, A	107
Owlflies: Ascalaphidae		L	A	108
Snakeflies: Raphidiidae			L, A	109

Note: L: larvae; A: adults.

Adults: Antlions

Larvae: Doodlebugs

Family: Myrmeleontidae
(mur-muh-lee-AHN-tih-dee)

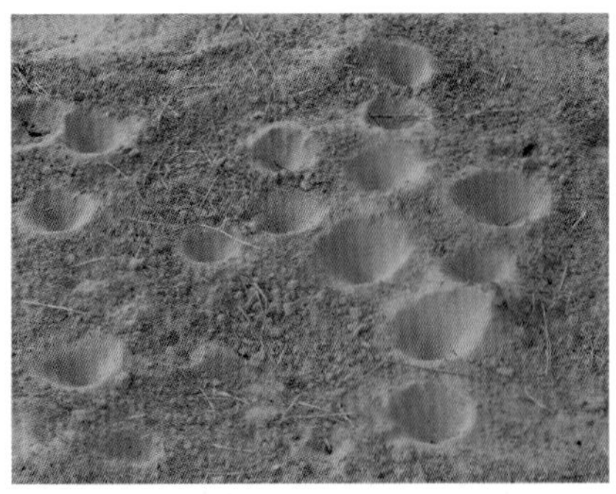
Antlion larvae burrows

Antlion larva

Characteristics:

SIZE: 2 to 4 in. (5.1 to 10.2 cm)
COLOR: **Brown**
ANTENNAE: **Threadlike with club at tip**
MOUTHPARTS: Chewing
EYES: Compound
WINGS: **Many veins and cross veins; clear or with irregular spots**
LEGS: Walking
MISCELLANEOUS: **Resemble damselflies; adults attracted to lights**
HABITAT:
Adults: Plants
Larvae: **Construct conical pit in loose, sandy areas; under debris**
FOOD:
Adults: Nectar, pollen; predaceous on small arthropods
Larvae: Predaceous on small insects and other arthropods

Antlion adult (by Gary Brooks)

Antlion adult

Antlions resemble damselflies with their long, slender abdomen and wings. Antlions, however, have noticeably longer antennae that are slightly clubbed at the tip. These insects are very weak fliers, are active at night, and are often attracted to lights.

The larval stage, known as "doodlebugs," is best known for the pits that they build and where they live. These pits are constructed of loose sand in protected areas. A small, hapless victim passing by will lose its footing and begin to slide down into the pit. As the prey slides within reach, the doodlebug will grab it with its large, sickle-shaped, hollow mandibles and suck the fluids of the victim's body.

Brown Lacewings

Family: Hemerobiidae
(hee-mehr-oh-BEE-ih-dee)

Brown lacewing larva with aphid prey

Characteristics:

SIZE: $\frac{3}{8}$ to $\frac{1}{2}$ in. (9.6 to 12.8 mm)
COLOR: **Brown**
ANTENNAE: Threadlike; not clubbed
MOUTHPARTS: Chewing
EYES: Compound
WINGS: **Many veins and cross veins; held rooflike when at rest**
LEGS: Walking
MISCELLANEOUS: **Resemble green lacewings; active at dusk and night**
Larvae: Similar to green lacewing larvae; more slender and smaller head
HABITAT:
- Adults: Wooded areas
- Larvae: Associated with food host

FOOD:
- Adults: Predaceous on small insects and other arthropods
- Larvae: Predaceous on small insects and other arthropods

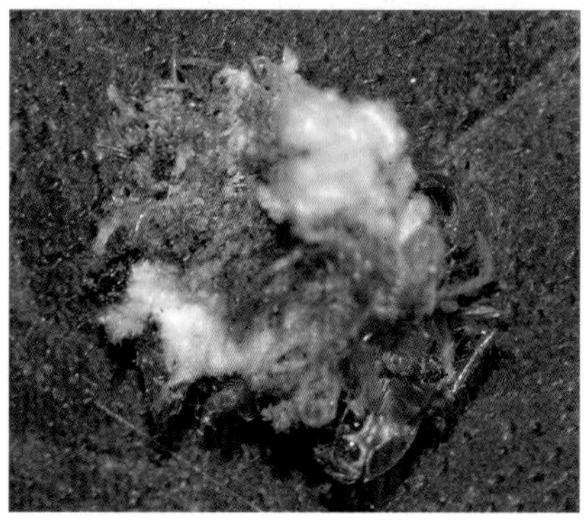

Trashbug larva

Brown lacewings are about one-half inch long with relatively large wings that have many cross veins and are held rooflike. These insects resemble green lacewings but differ in color and are smaller. Brown lacewings are less common and prefer cooler weather than green lacewings and are more common in wooded than open areas.

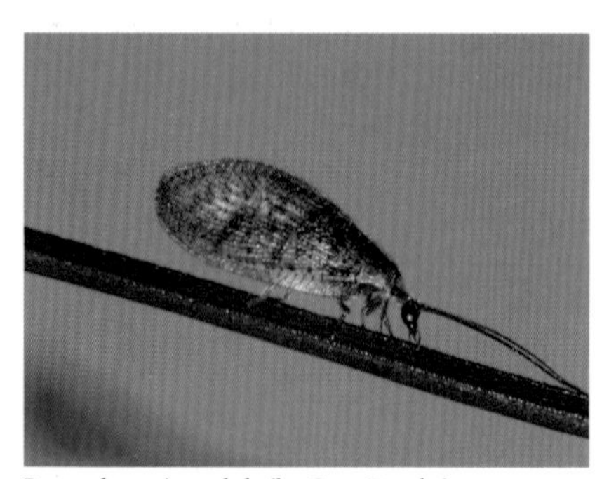

Brown lacewing adult (by Gary Brooks)

Both the adult and larval stages feed on small arthropods, particularly aphids. Adult female brown lacewings lay their eggs on leaves, twigs, or bark but not on a stalk as do the green lacewings.

The larvae of some species are known as "trashbugs" because they cover their bodies with the skins of their victims.

Adults: Dobsonflies

Larvae: Hellgrammites

Family: Corydalidae
(kor-ee-DAL-ih-dee)

Characteristics:

SIZE: **2 to 3 in. (5.0 to 7.6 cm)**
COLOR: **Tan**
ANTENNAE: Threadlike
MOUTHPARTS: Chewing; **males greatly elongated and sickle shaped; females short and stout**
EYES: Compound; three ocelli
WINGS: Many veins and cross veins; held rooflike over body when at rest
LEGS: Walking
MISCELLANEOUS: **Soft bodied**
Adults: Nocturnal and attracted to lights
Larvae: Can grow to over 2 in.; commonly sought after as fish bait
HABITAT:
Adults: Near water
Larvae: Aquatic; under stones in flowing streams
FOOD:
Adults: Probably do not feed
Larvae: Predaceous on aquatic insects and other small arthropods

Female dobsonfly (by Curtis Williams)

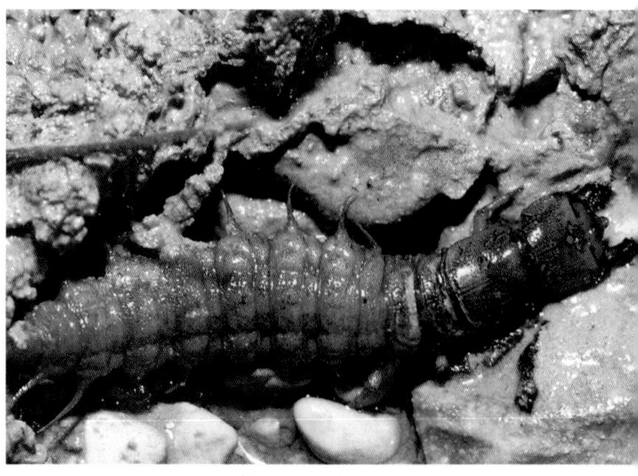

Hellgrammite

Dobsonflies are among Texas' largest insects with some individuals more than three inches long. They are weak fliers and often attracted to lights. Male dobsonflies can be easily identified by their extremely long mandibles that resemble fangs. Although fierce looking, they are harmless. Some authorities place corydalids in a separate order, Megaloptera.

The larvae, referred to as "hellgrammites," are aquatic and typically can be found under stones in moving water. Hellgrammites are predaceous on insects and other small aquatic animals. They can grow to over two inches in length and are prized as catfish bait by many who fish with trotlines. The larvae construct cells under rocks on the shore to pupate and overwinter in this stage. The entire life cycle may require up to three years to complete.

Adults: Green Lacewings

Larvae: Aphidlions

Family: Chrysopidae
(kry-SAH-pih-dee)

Characteristics:

SIZE: $\frac{1}{2}$ to 1 in. (1.3 to 2.5 cm)
COLOR: **Lime green**
ANTENNAE: Threadlike
MOUTHPARTS: Chewing
EYES: Compound; **typically golden**
WINGS: Many veins and cross veins; held rooflike when at rest
LEGS: Walking
Larvae: **Spindle shaped; long, sickle-shaped mouthparts**
HABITAT:
Adults: Plants
Larvae: Associated with food host

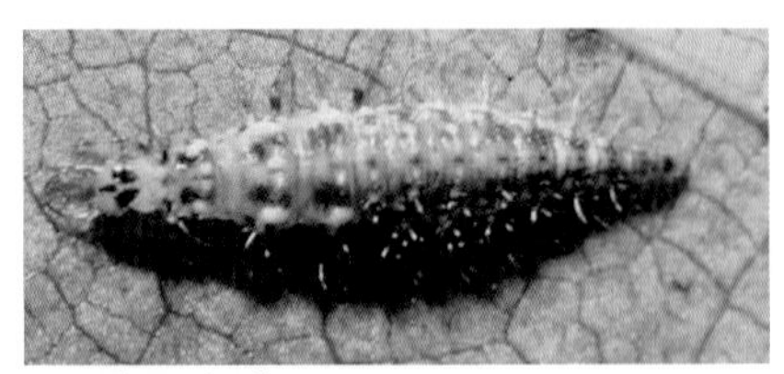

Green lacewing larva

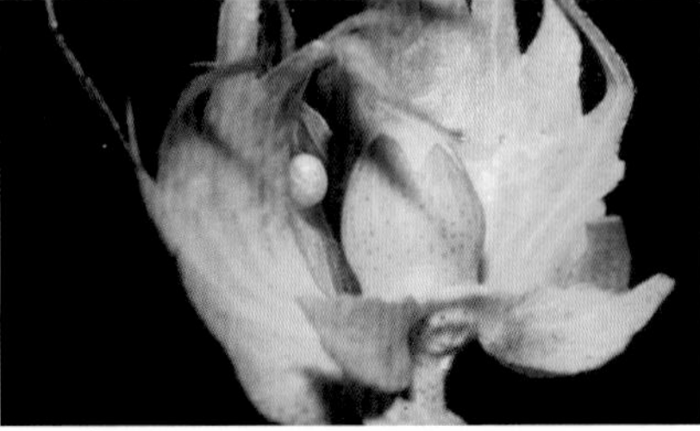

Green lacewing pupa

Green lacewing adult

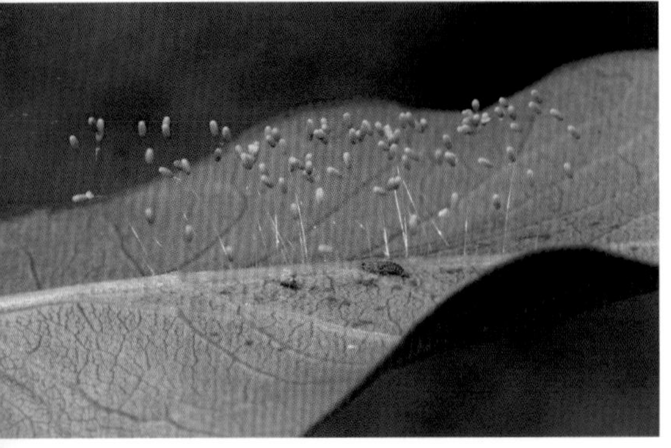

Green lacewing eggs (by Gary Brooks)

FOOD:
Adults: Predaceous on small arthropods; nectar
Larvae: Predaceous on small arthropods

Green lacewings are probably the most common and economically beneficial member of the order Neuroptera. Adult green lacewings are about two-thirds inch long and lime green with golden eyes.

The adult female green lacewing lays her eggs on a stalk singly or in clusters, which may be confused with fungal growth. The larvae, known as "aphidlions," are shaped like tiny alligators with long, sickle-shaped mouthparts. Aphidlions use these hollow mouthparts to impale aphids and other small arthropods and suck the juices from their bodies.

After the larval stage is complete, the mature larva spins an egg-shaped cocoon and attaches itself to a nearby structure. These are also sometimes misidentified as spider eggs.

Aphidlions are considered one of the more important predators of aphids and can be purchased for release in home gardens.

Mantidflies

Family: Mantispidae
(man-TIHS-pih-dee)

Characteristics:

SIZE: $\frac{1}{2}$ to 1 in. (1.3 to 2.5 cm)
COLOR: Brown
ANTENNAE: Threadlike
MOUTHPARTS: Chewing
EYES: Compound
WINGS: Many veins and cross veins; **held rooflike when at rest**
MISCELLANEOUS: **Prothorax elongated; resemble praying mantids**
LEGS: **Forelegs modified for grasping and separated from back legs;** mid- and hind legs for walking
HABITAT:
 Adults: Terrestrial
 Larvae: Host habitat
FOOD:
 Adults: Predaceous on small insects
 Larvae: Parasitoids of spider eggs or wasp larvae

Most mantidflies, as the name implies, resemble praying mantids. Both have an elongated prothorax and grasping forelegs, which are used to catch and hold their prey. Mantidflies can be distinguished by their smaller size and the way they hold their wings rooflike over their backs. Some species resemble wasps.

Female mantidflies lay their eggs on a short stalk in clusters of several hundred. After hatching, the larvae must locate their food source, which is either spider eggs or Hymenoptera larvae. The young mantispid larva will either seek the actual food source or catch a ride on the adult spider or wasp and stay with it until the host deposits her eggs.

Once the food source is located, the mantispid larva will feed on the developing spider eggs or Hymenoptera larva. After completing development, the mature larva pupates inside the host egg case or larval cell.

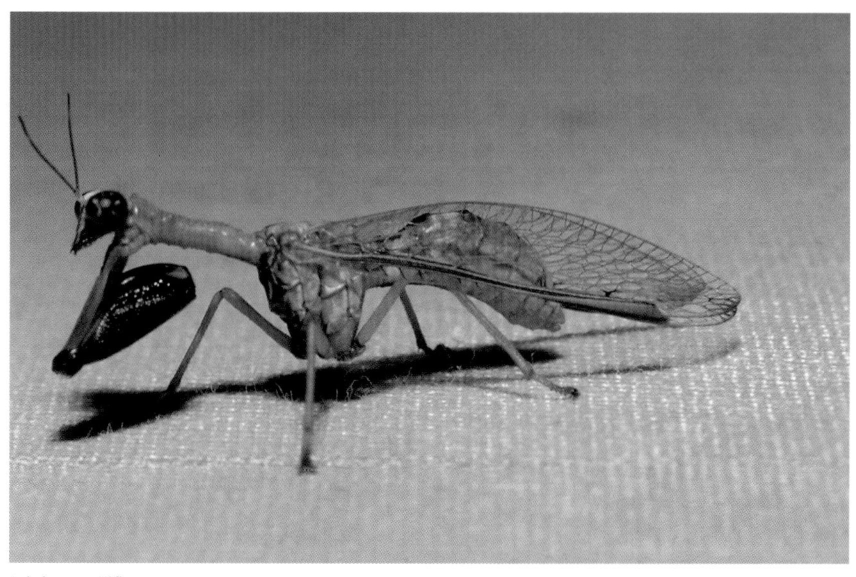

Adult mantidfly

Owlflies

Family: Ascalaphidae
(as-kuh-LAY-fih-dee)

Characteristics:

SIZE: 1 to 2 in. (2.5 to 5.1 cm) (excluding antennae)
COLOR: **Brownish gray**
ANTENNAE: **Long; threadlike; end with a knob**
MOUTHPARTS: Chewing
EYES: Compound, large, and bulging
WINGS: Many veins and cross veins; held rooflike when at rest
LEGS: Walking
MISCELLANEOUS: **Abdomen resembles twig and is held perpendicular to resting surface;** adults nocturnal
HABITAT:
Adults: Trees and shrubs
Larvae: Ground litter
FOOD:
Adults and Larvae: Predaceous on small arthropods

Owlflies are nocturnal insects seldom seen by the casual observer and then often mistaken for dragonflies. These insects can be identified both by their very long antennae that terminate in a club and by the way they hold their wings rooflike instead of outstretched as dragonflies do. When at rest, the adult owlfly "hides" by placing its head and antennae parallel to the branch and raising its tail, which resembles a twig.

The female owlfly lays her eggs on twigs, and the newly hatched larvae crawl to the ground. The larvae resemble antlions but do not build pits; rather, they conceal themselves with ground litter. From these protective hideouts, the ascalaphid larvae can ambush passing small arthropods.

Adult owlfly

Snakeflies

Family: Raphidiidae
(ruh-FIH-dih-ih-dee)

Characteristics:
SIZE: $\frac{3}{8}$ to $\frac{3}{4}$ in. (9.6 to 19.2 mm)
COLOR: Dark
ANTENNAE: Threadlike
MOUTHPARTS: Chewing
EYES: Compound
WINGS: Many veins and cross veins; held rooflike when at rest
LEGS: Walking; **forelegs borne at posterior end of elongated prothorax**
MISCELLANEOUS: **Prothorax elongated; appear to have a long neck; female ovipositor generally long**
HABITAT:
Adults: Trees
Larvae: Trees; under bark
FOOD:
Adults and Larvae: Predaceous on small arthropods

Snakeflies are uncommon insects whose elongated prothorax and loosely joined head give them a snakelike appearance. They somewhat resemble mantispids, but their forelegs are not grasping and are located on the posterior end of the prothorax. Some entomologists place these in a separate order, Raphidioptera.

Adult snakeflies stalk their prey in a manner similar to snakes. When an insect comes within reach, the snakefly will strike, catching the victim with its jaws.

The female snakefly has a long ovipositor that she uses to insert eggs into the bark of trees. The larvae reach one-half to three-fourths inch and feed on small tree-dwelling arthropods.

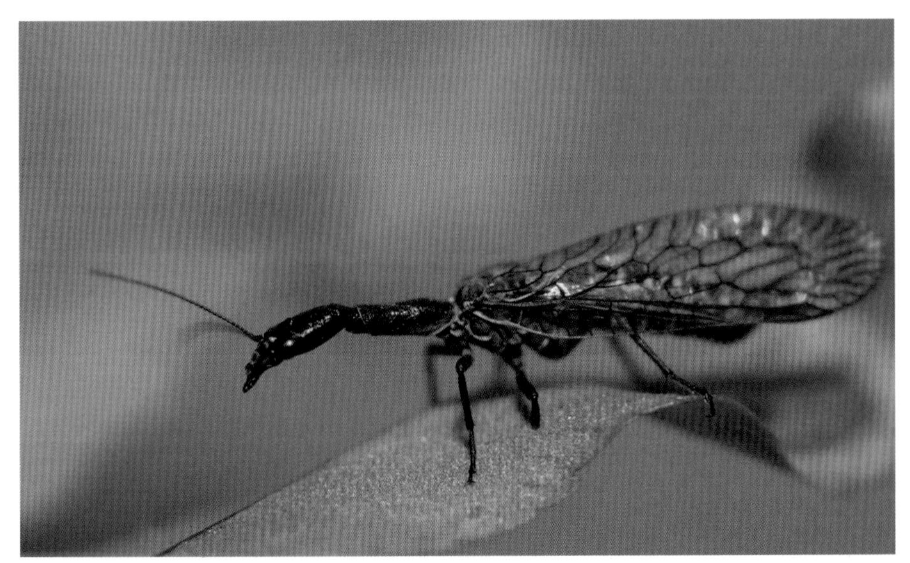

Adult snakefly (by Gary Brooks)

Beetles and Weevils

Order: Coleoptera
(koh-lee-AHP-tur-uh)
coleo, sheath; *ptera,* wing

Characteristics:

SIZE: Variable
ANTENNAE: Variable
MOUTHPARTS: Chewing
EYES: Compound
WINGS: Two pairs; **first pair (called elytra) hard and meet in a straight line down the back**
LEGS: Generally for walking; some adapted for swimming
FOOD: Plant feeders; predators; scavengers; very few parasitic
METAMORPHOSIS: Complete (egg, larva, pupa, adult)

Beetles are the largest order of insects. Equally amazing is the diversity of beetles' habitats. Many are found on and eat plants, while others are valuable predators of plant-eating insects. Some live in various aquatic environments. Others are important in the decomposition of dead plants and animals.

Members of the order Coleoptera are generally easy to recognize by the thickened forewings that meet in a straight line down their back. These wings, called elytra, are not used for flight but rather as protection for the membranous rear flight wings. They have chewing mouthparts that, in some species, are elongated and appear to be piercing/sucking, but in fact are toothed at the tip.

Common families of the order Coleoptera and their preferred habitats

Family	*Aquatic*	*Decomposing materials*	*Plants*	*Ground*	*Page*
Bess beetles: Passalidae		L, A			112
Blister beetles: Meloidae			A	L	113
Carrion beetles: Silphidae		L, A			114
Click beetles (wireworms): Elateridae		L, A	L, A	L	115
Darkling beetles: Tenebrionidae		L, A	L, A	L, A	116
Fireflies: Lampyridae			A	L	117
Ground beetles: Carabidae				L, A	118
June beetles, scarab beetles, dung beetles (white grubs): Scarabaeidae		L, A	L, A	L, A	119
Ladybird beetles: Coccinellidae			L, A		120
Leaf beetles: Chrysomelidae			L, A		121
Long-horned beetles (round-headed wood borers): Cerambycidae			L, A		122
Metallic wood-boring beetles (flat-headed wood borers): Buprestidae			L, A		123
Net-winged beetles: Lycidae			L, A		124
Predaceous diving beetles: Dytiscidae	L, A				125
Rove beetles: Staphylinidae		L, A			126
Skin beetles: Dermestidae		L, A			127
Soldier beetles: Cantharidae			L, A	L	128
Tiger beetles: Cicindelidae				L, A	129
Tumbling flower beetles: Mordellidae			A	L	130
Twig borers: Bostrichidae			L, A		131
Water scavenger beetles: Hydrophilidae	L, A				132
Weevils, snout beetles: Curculionidae			L, A		133
Whirligig beetles: Gyrinidae	L, A				134

Note: L: larvae; A: adults.

Bess Beetles

Family: Passalidae
(puh-SAL-ih-dee)

Characteristics:

SIZE: $1\frac{1}{8}$ to $1\frac{5}{8}$ in. (2.9 to 4.1 cm)
SHAPE: **Elongated; parallel sided**
COLOR: **Typically black or brown and shiny**
ANTENNAE: Clubbed; 3-segmented
MOUTHPARTS: Chewing
EYES: Compound
WINGS: **Grooved**
LEGS: Walking; tarsal formula 5-5-5
MISCELLANEOUS: **Pronotum with grooved midline; short horn on the head**
HABITAT:
Adults and Larvae: Rotting wood
FOOD:
Adults: Decaying wood
Larvae: Decaying wood with microorganisms

Bess beetles are found in decomposing trees and form a somewhat social network of galleries. They communicate by rubbing their abdomen against their wings to make a squeaking sound. The adults construct tunnels lined with finely chewed wood pulp mixed with secretions from their salivary glands. This predigested material serves as the primary food source for the developing larvae.

Adult bess beetle

Blister Beetles

Family: Meloidae
(muh-LOY-dee)

Characteristics:

SIZE: $\frac{1}{4}$ to $1\frac{1}{4}$ in. (6.4 to 31.8 mm)

SHAPE: **Mostly narrow and elongated;** a few roundish

COLOR: **Extremely variable in color and size; solid, striped, or spotted**

ANTENNAE: Threadlike; 11-, rarely 8- or 9-segmented

MOUTHPARTS: Chewing

EYES: Compound; **partially wrap around base of antennae**

WINGS: **Tip of wings rounded exposing tip of abdomen;** soft and flexible; short in some species

LEGS: Walking; long and slender; tarsal formula 5-5-4

MISCELLANEOUS: **Neck narrower than head and body**

HABITAT:
- Adults: Plants
- Larvae: Same habitat as host

FOOD:
- Adults: Plant foliage or flowers; nectar; pollen
- Larvae: Predaceous on grasshopper eggs and larvae of social bees and wasps

Blister beetle (note notched eyes and narrow neck)

Adult blister beetle

Adult blister beetle (by Gary Brooks)

The adult blister beetle is considered a pest because it feeds on and can defoliate many plants, including crops and ornamentals. The larvae of some species, however, are predaceous on grasshopper eggs and are considered to be a valuable natural control agent. The larvae of other species lie in wait for a passing wild bee or wasp and grab hold. Then when the bee or wasp unknowingly takes the larva back to its nest, the larva feeds on its host's eggs and larvae.

Blister beetles undergo a complex developmental process known as hypermetamorphosis. Typically there are four larval stages, each of which has a distinctively different appearance.

Many species of blister beetles produce a defensive compound known as cantharidin, which can cause blisters and can be quite toxic to horses that eat hay containing these insects.

Carrion Beetles

Family: Silphidae
(SIHL-fih-dee)

Characteristics:

SIZE: $\frac{1}{3}$ to $1\frac{1}{4}$ in. (8.5 to 31.8 mm)

SHAPE: **Wings broadest at rear;** somewhat flattened

COLOR: **Usually black with red, orange, or yellow markings on wings**

ANTENNAE: Clubbed; 10- or 11-segmented

MOUTHPARTS: Chewing

EYES: Compound

WINGS: **Often short and square, exposing tip of abdomen**

LEGS: Walking; tarsal formula 5-5-5

MISCELLANEOUS: **Pronotum wider than head**

HABITAT:

Adults and Larvae: **Decaying animal material**

FOOD:

Adults and Larvae: Carrion; predaceous on other insects

Adult carrion beetle

Adult carrion beetle

Carrion beetles are generally black with bright red, orange, or yellow markings on the wings. Their wings are short and blunt, exposing the rear portion of the abdomen.

Silphids possess a keen sense of smell and can locate dead animals. Both the adults and larvae are associated with decomposing animals.

The adult carrion beetles of one group are called "burying beetles." These beetles dig beneath the corpse until the corpse slowly sinks and is completely below ground. Because flies typically will not lay eggs on buried corpses, young carrion beetle larvae can feed freely without competition from maggots. Any maggot that does get to the corpse is eaten by the adult beetles.

Adults: Click Beetles

Larvae: Wireworms

Family: Elateridae
(ee-luh-TUR-ih-dee)

Characteristics:

SIZE: $\frac{1}{8}$ to $1\frac{1}{2}$ in. (3.2 to 38.1 mm)
SHAPE: **Body elongated with both ends rounded; pronotum U-shaped with sharp points**
COLOR: Usually brown or black; sometimes with markings
ANTENNAE: **Usually saw-toothed;** 11-segmented
MOUTHPARTS: Chewing
EYES: Compound
WINGS: **Rounded at tip;** usually striated; parallel sided
LEGS: Walking; tarsal formula 5-5-5
MISCELLANEOUS: **Prothorax loosely joined to mesothorax; body can bend in the middle; some species possess eyespots on prothorax**
HABITAT:
Adults: Flowers; foliage; rotting logs
Larvae: Soil; rotting logs
FOOD:
Adults: Plant foliage
Larvae: Plant seeds and roots; predaceous on other insects

Adult eyed elater, *Alaus oculatus*

Click beetles have the unique ability to flip over when they are placed or land on their back. The first and second thoracic segments are loosely joined (unlike in most insects, where the segments are rigidly joined), allowing the beetle to "bend" in the middle. A finger-like projection on the underside of the second thoracic segment fits into a slot located on the first segment.

When upside down, the click beetle arches its back so only the tips of its head and abdomen touch ground. With a sudden snap, the beetle propels itself into the air, sometimes over an inch high, making a clicking sound in the process.

The larvae are long and cylindrical. Their unusually tough skin gives them the name "wireworms." Wireworms typically live in the soil and feed on plant seeds, roots, and other underground vegetation. Some species can cause economic loss to agricultural crops, while other species live in decomposing logs and feed on other small insects and arthropods.

Adult click beetle

Wireworm

Darkling Beetles

Family: Tenebrionidae
(tehn-ee-bree-AHN-ih-dee)

Characteristics

SIZE: $\frac{1}{8}$ to $1\frac{1}{2}$ in. (3.2 to 38.1 mm)
SHAPE: Variable; **some resemble ground beetles**
COLOR: **Mostly dull black or brown**
ANTENNAE: Threadlike, beadlike, or clubbed; inserted beneath front of head; 11-segmented
MOUTHPARTS: Chewing
EYES: Compound
WINGS: **Typically striated; wrap around sides of abdomen**
LEGS: Walking; tarsal formula 5-5-4
MISCELLANEOUS: **Some species stand on head when disturbed;** adults nocturnal
HABITAT:
Adults: **Many species prefer arid regions;** ground; decomposing logs
Larvae: Decomposing materials; stored cereal grains
FOOD:
Adults: Decaying vegetable matter; stored products
Larvae: Decaying matter; fungi; stored products

Tenebrionidae is a large group of beetles that prefer arid regions. Many species resemble ground beetles since they are typically dull black or brown and often found on the ground. Upon close inspection, however, the wings of darkling beetles wrap around past the midpoint on the sides. Tenebrionids also have a 5-5-4 tarsal formula, whereas carabids have a 5-5-5 tarsal formula.

Some species, particularly those of the genus *Eleodes,* stand on their head when disturbed and emit a foul-smelling defensive substance. The ironclad beetle, *Zopherus nodulosus,* is common in wooded regions of Texas and gets its name from its very hard exoskeleton. Ironclad beetles are sometimes placed in a separate family, Zopheridae.

The larvae of many tenebrionids, like the adults, prefer arid habitats. The confused flour beetle, *Tribolium confusum,* can be a serious pest in flour. Yellow mealworm, *Tenebrio molitor,* larvae are often reared as food for small captive animals and pets.

Ironclad beetle, *Zopherus nodulosus*

Darkling beetle

Fireflies, Lightningbugs

Family: Lampyridae
(lam-PEER-ih-dee)

Characteristics:

SIZE: $\frac{1}{4}$ to $\frac{7}{8}$ in. (6.4 to 22.4 mm)
SHAPE: Cylindrical; **head covered by prothorax**
COLOR: Typically brownish to olive color; often marked with black; **luminous abdominal segments lighter colored**
ANTENNAE: Sawlike or threadlike; 11-segmented
MOUTHPARTS: Chewing
EYES: Compound
WINGS: **Soft; rounded at tip**
LEGS: Walking; tarsal formula 5-5-5
MISCELLANEOUS: Females winged, short winged, or wingless
HABITAT:
Adults: Damp, wooded areas
Larvae: Under stones; leaf litter; other damp areas
FOOD:
Adults: Most do not eat; predaceous on other insects
Larvae: Predaceous on snails and other soft-bodied organisms

Firefly (note two-colored abdomen)

Who has not marveled at the sight of hundreds of lightningbugs flashing in a low-lying wooded area? This spectacular light show is the result of the fireflies signaling potential mates. Each species has its own inherent, specific sequence of flashes.

The light produced is a result of a chemical reaction involving the compound luciferin. This truly amazing chemical reaction produces nearly 100 percent light and yet releases very little heat.

The adults of some species do not eat; the adults of other species, and all larvae, are predaceous. The females of some species can copy the flash pattern of other species, and a hapless male that responds to these copied flashes may find a hungry rather than amorous female.

Ground Beetles

Family: Carabidae
(kuh-RAB-ih-dee)

Characteristics:

SIZE: ¼ to 1¾ in. (6.4 to 44.5 mm)
SHAPE: Variable; usually parallel sided; elongated and somewhat flattened
COLOR: **Generally black and shiny; some species brightly colored**
ANTENNAE: Threadlike; inserted between eyes and base of mandibles; 11-segmented
MOUTHPARTS: Chewing; sickle shaped
EYES: **Compound eyes narrower than pronotum**
WINGS: **Typically striated**
LEGS: Long and slender; tarsal formula 5-5-5
MISCELLANEOUS: Kidney-shaped structure at base of hind coxae; **adults nocturnal; often attracted to lights**

Ground beetle

Ground beetle larva

Caterpillar hunter, *Calosoma scrutator*

HABITAT:
Adults: **Generally on ground, under logs, stones, etc.**
Larvae: Generally on ground, under logs, stones, etc.
FOOD:
Adults: Mostly omnivorous; predaceous on other insects; decomposing materials; rarely, seeds
Larvae: Predaceous on other small arthropods

Among the more commonly encountered beetles in Texas, ground beetles are generally nocturnal and often attracted to lights. During the day, they remain hidden under any object that offers protection from predators. Most species are dark and drab in color with striated wings, although a few species are quite colorful. The caterpillar hunters, *Calosoma scrutator,* are large, bright metallic green beetles that are fairly common following outbreaks of other insects such as crickets and moths.

Most species are predaceous on a wide variety of hosts but will also eat dead insects. Members of the genus *Calosoma* feed on caterpillars, and *Carabus* feed on earthworms and snails.

June Beetles (Bugs), Scarab Beetles, May Beetles, Dung Beetles, Tumblebugs

Larvae: White Grubs

Family: Scarabaeidae
(skair-uh-BEE-ih-dee)

Green June beetle, *Cotinis* spp.

Scarabid larva

Characteristics:

SIZE: $\frac{1}{8}$ to $2\frac{1}{2}$ in. (3.2 to 63.5 mm)
SHAPE: **Distinctive oval shape; convex**
COLOR: **Generally brown or black but sometimes brightly colored**
ANTENNAE: **Flag shaped;** 8- to 11-segmented
MOUTHPARTS: Chewing
EYES: Compound
WINGS: Convex; sometimes exposing tip of abdomen
LEGS: Walking; forelegs often adapted for digging; tarsal formula 5-5-5
MISCELLANEOUS: **Males sometimes with horn**
HABITAT:
Adults: Decomposing materials; plants
Larvae: **Soil; decomposing materials**
FOOD:
Adults: Plant roots and leaves; dung; carrion; fruit; pollen
Larvae: Decomposing material; plant roots

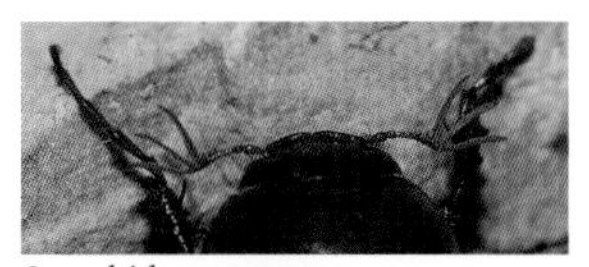
Scarabid antennae

Dung beetle

Scarab beetles are a large and very common group of beetles easily recognized by their oval body and flag-shaped antennae. Most are dull colored, but some are metallic and quite attractive. Commonly called "white grubs," the typically white larvae are generally C-shaped and possess well-developed legs.

Scarabs occupy a wide variety of habitats. The larvae of some species, such as May beetles, *Phyllophaga* spp., feed on plant roots and can be serious pests of agricultural crops and turfgrass. Other species are important in decomposing animal wastes and, in the process, destroying the habitat of flies and other manure-dwelling insects. Adults of dung-feeding tumblebugs chew off a piece of dung and roll it into a ball. The ball is buried in the soil; the larvae live and feed in this protected environment. Skin beetles, *Trox* spp., are the last to leave an animal carcass, feeding on the remaining hide and hair. Some authorities place skin beetles in a separate family, Trogidae.

Ladybird Beetles, Ladybugs

Family: Coccinellidae
(kahks-ih-NEHL-ih-dee)

Characteristics:

SIZE: $\frac{1}{4}$ to $\frac{3}{8}$ in. (6.4 to 9.6 mm)

SHAPE: **Hemispherical; head concealed under the pronotum**

COLOR: **Solid color or spotted; seldom striped**

ANTENNAE: Clubbed; 8- to 11-segmented with 3- to 6-segmented club

MOUTHPARTS: Chewing

EYES: Compound

WINGS: **Convex**

LEGS: Walking; tarsal formula appears 3-3-3, actually 4-4-4

HABITAT:

Adults and Larvae: Typically on plants associated with prey

FOOD:

Adults: Predaceous on small insects; a few plant-feeding exceptions

Larvae: Predaceous on small insects; a few feed on plants

One of the most recognizable groups of insects, ladybird beetles are well known for their role as predators. Both the adult and larval stages feed on small insects, particularly aphids, and are often purchased and released with hopes of controlling arthropod pests on a variety of plants. A few species, including the Mexican bean beetle, *Epilachna varivestis,* are plant feeders and can cause considerable damage.

The adult females lay clusters of football-shaped yellow eggs. The larvae are typically spindle shaped, quite active, and commonly found in aphid colonies. The larvae of other species (particularly *Scymnus* spp.) produce a waxy substance that covers their bodies. These resemble and are sometimes confused with mealybugs.

Ladybird beetle

Ladybird beetle

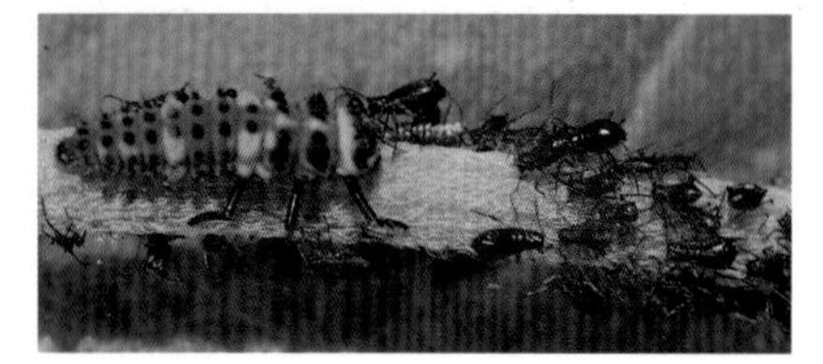

Ladybird beetle larva with aphids

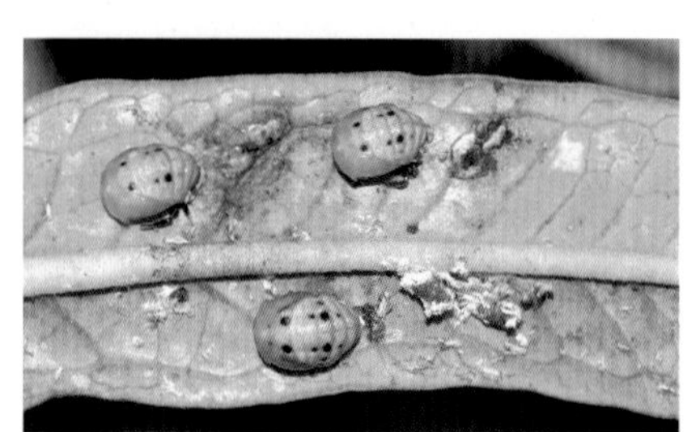

Ladybird beetle pupae

Leaf Beetles

Family: Chrysomelidae
(kry-soh-MEHL-ih-dee)

Characteristics:

SIZE: $\frac{1}{16}$ to $\frac{1}{2}$ in. (1.6 to 12.8 mm)
SHAPE: **Very diverse**
COLOR: **Solid color, striped, or spotted; often brightly colored**
ANTENNAE: Threadlike; less than half the length of the body; 11-segmented
MOUTHPARTS: Chewing
EYES: Compound
WINGS: Typically well developed
LEGS: Walking; hind legs often enlarged for jumping; tarsal formula apparently 4-4-4, actually 5-5-5
MISCELLANEOUS: Sometimes resemble lady beetles
HABITAT:
Adults and Larvae: Plants
FOOD:
Adults: Flowers and leaves
Larvae: Mostly leaves and roots; some are leaf miners; some are stem borers

Adult Colorado potato beetle, *Leptinotarsa decemlineata*

Colorado potato beetle larva, *Leptinotarsa decemlineata*

Flea beetles (by Gary Brooks)

Twelve-spotted cucumber beetle, *Diabrotica undecimpunctata howardi*

Members of this large, diverse group are sometimes difficult to identify. Many species resemble ladybird beetles but are more oval, while others are similar to long-horned beetles but have shorter antennae.

All species are plant feeders, and many are important pests. Among the more notorious species are the spotted cucumber beetle (larva, corn rootworm), *Diabrotica undecimpunctata howardi;* Colorado potato beetle, *Leptinotarsa decemlineata;* and the elm leaf beetle, *Xanthogaleruca luteola.* Most species restrict their feeding activity to closely related plant species, but a few have a wide host range.

A few species are important beneficial insects feeding on weeds. The klamathweed beetle, *Chrysolina quadrigemina,* was introduced into California to control klamathweed, *Hypericum perforatum,* and has proven to be remarkably successful.

Adults: Long-horned Beetles

Larva: Round-headed Woodborers

Family: Cerambycidae
(sur-uhm-BIHS-ih-dee)

Characteristics:

SIZE: $\frac{1}{4}$ to $2\frac{3}{8}$ in. (6.4 to 60.3 mm)
SHAPE: **Elongated and cylindrical**
COLOR: Variable; often very colorful
ANTENNAE: Threadlike; **almost always at least half the length of the body, often longer;** inserted so that the base is partially surrounded by the eyes; 11- to 25+-segmented
MOUTHPARTS: Chewing
EYES: **Compound eyes usually notched by antennae**
WINGS: **Usually wider than pronotum**
LEGS: Walking; tarsal formula appears 4-4-4, actually 5-5-5
HABITAT:
  Adults: Plants, typically flowers
  Larvae: Deciduous trees
FOOD:
  Adults: **Pollen;** flowers; leaves; wood
  Larvae: **Deciduous trees; sapwood and heartwood**

Long-horned beetle

Locust borer, *Megacyllene robiniae*

Most long-horned beetles are recognized by their antennae, which are typically at least half the length of their body but may be considerably longer in some species. The male's antennae are generally longer than the female's.

Many long-horned beetles are very attractive and are among the most sought after insect families by collectors.
The larvae, known as round-headed borers, can be serious pests of shade trees, although most attack damaged, dying, or freshly cut trees. Adult long-horned beetle females lay eggs in cracks in wood, and the white, cylindrical larvae burrow into the tree, leaving a circular hole. The life cycle may take up to three years to complete.

The black-and-yellow locust borer, *Megacyllene robiniae,* resembles a wasp and attacks freshly cut firewood. It is common to hear the larvae eating inside firewood and see the sawdust as evidence of their presence.

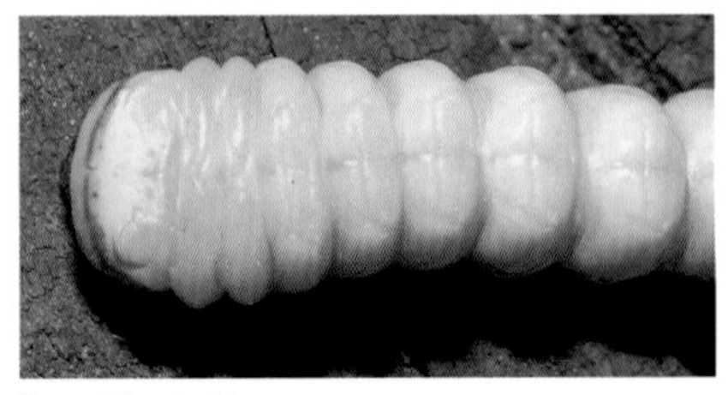

Round-headed borer

Adults: Metallic Wood-Boring Beetles

Larvae: Flat-headed Borers

Family: Buprestidae
(byoo-PREHS-tih-dee)

Characteristics:

SIZE: $\frac{1}{3}$ to $1\frac{1}{4}$ in. (8.5 to 31.8 mm)
SHAPE: **Bullet shaped**
COLOR: **Often metallic; sometimes spotted**
ANTENNAE: Generally saw-toothed, sometimes threadlike
MOUTHPARTS: Chewing
EYES: Compound
WINGS: **Pointed at tip**
LEGS: Walking; tarsal formula 5-5-5
MISCELLANEOUS: **Head sunken into prothorax; body surface often pitted or grooved**
HABITAT:
Adults: On or near trees; typically on the sunny side
Larvae: **Within trees**
FOOD:
Adults: Pollen and nectar; foliage
Larvae: Most are wood borers

Metallic wood borer

Metallic wood borer

Metallic wood-boring beetles superficially resemble click beetles, but their bodies are not flexible and are often metallic colored. Adult buprestids are found in sunny areas on flowers, where they feed on pollen. A few feed on foliage or twigs.

Adult female buprestids lay their eggs in the cracks of trees. The larvae are called "flat-headed wood borers" because the first segment behind their head is large and flattened. These larvae enter the tree trunk at an angle and produce winding, flattened tunnels. They prefer weakened or damaged trees. A few species are gall formers.

Metallic wood borer

Flat-headed borer (by Gary Brooks)

Net-winged Beetles

Family: Lycidae (LY-sih-dee)

Characteristics:

SIZE: $\frac{1}{4}$ to $\frac{3}{4}$ in. (6.4 to 19.2 mm)

SHAPE: **Widen toward rear; fan shaped**

COLOR: **Usually black with red or orange markings**

ANTENNAE: Long, flattened; generally saw-toothed or threadlike; 11-segmented

MOUTHPARTS: Chewing

EYES: Compound

WINGS: **Broaden toward the rear, appearing fan shaped; netlike veins**

LEGS: Walking; tarsal formula 5-5-5

MISCELLANEOUS: **Head concealed by pronotum;** adults diurnal

HABITAT:

Adults: Dense foliage; wooded areas

Larvae: Under bark

FOOD:

Adults: Decaying plant juices; predaceous on other small arthropods

Larvae: Soft, rotting wood; predaceous on other small arthropods

Net-winged beetles can be recognized by their bright color and rough-textured, fan-shaped wings. They are found in dense, wooded areas.

Adult lycids are distasteful, and their bright coloration serves as a warning to would-be predators that the insect contains toxic chemicals or possesses a sting. Known as aposematic coloration, this is a common warning system among insects.

The larvae of net-winged beetles are found under tree bark and may feed on soft rotting wood, slime molds, or other small arthropods.

Net-winged beetle

Adults: Predaceous Diving Beetles

Larvae: Water Tigers

Family: Dytiscidae
(dy-TIHS-ih-dee)

Characteristics:

SIZE: $\frac{1}{2}$ to $1\frac{1}{2}$ in. (12.8 to 38.1 mm)

SHAPE: **Body smooth and oval; convex and streamlined**

COLOR: **Black or brown and often with indistinct yellow markings**

ANTENNAE: **Threadlike; long**

MOUTHPARTS: Chewing; **palps short**

EYES: Compound

WINGS: Hard; meet in a straight line down back

LEGS: **Hind legs modified for swimming, flattened to form paddles, and fringed with hair**

MISCELLANEOUS: **When swimming, hind legs move together like oars**

HABITAT:

Adults: **Shallow ponds and quiet streams**

Larvae: Shallow ponds and quiet streams

FOOD:

Adults and Larvae: Predaceous on small aquatic arthropods

Predaceous diving beetles are well suited for aquatic life. The adults possess large, flattened hind legs fringed with long hairs. Unlike most aquatic insects, these beetles stroke both legs at the same time when swimming. They possess a breathing apparatus at the end of their abdomen that allows them to keep their head below the surface when coming up for air. Before submerging, they collect a pocket of air under their front pair of wings.

The larvae are called "water tigers" because of their sickle-shaped jaws. Both the adults and larvae are predaceous on other insects and even small fish.

Predaceous diving beetle (by Stephen Kimbell)

Rove Beetles

Family: Staphylinidae
(staf-ih-LIHN-ih-dee)

Characteristics:

SIZE: $\frac{1}{8}$ to 1 in. (3.2 to 25.4 mm)

SHAPE: **Body narrow and parallel sided;** pronotum often larger than head

COLOR: Brown or black

ANTENNAE: Variable; usually thread-like, some clubbed

MOUTHPARTS: Chewing

EYES: Compound

WINGS: **Forewings short, exposing most of the abdomen; hind wings well developed and folded under forewings when at rest**

LEGS: Walking; tarsal formula usually 5-5-5

MISCELLANEOUS: **When running, may hold tip of abdomen up;** commonly attracted to light

Rove beetle

Rove beetle

HABITAT:

Adults: Variable; typically decaying material; ant or termite nests

Larvae: Moist areas, typically carrion

FOOD:

Adults: Predaceous on other small arthropods; carrion; fungi; a few parasitic on other insects

Larvae: Predaceous on other small arthropods

The rove beetles are the second-largest insect family, following the weevils and snout beetles. They are easily recognized by their short forewings, so most of the abdomen is exposed. The second pair of wings is intricately folded under the forewings when at rest. Rove beetles may be confused with earwigs but lack the well-defined pinchers at the tip of the abdomen. Many are attracted to lights.

Rove beetles are quick runners and excellent fliers. When alarmed, they will often run with their flexible abdomen pointed upward, as a scorpion does. However, they cannot sting. The adults and larvae are mostly predators of insects found in decomposing materials, such as fly larvae, and are considered beneficial. Other species are scavengers.

Skin Beetles, Larder Beetles, Carpet Beetles

Family: Dermestidae
(dur-MEHS-tih-dee)

Characteristics:

SIZE: $\frac{1}{16}$ to $\frac{1}{2}$ in. (1.6 to 12.8 mm)
SHAPE: **Oval or circular**
COLOR: **Typically dark; often with distinct patterns**
ANTENNAE: **Clubbed and fitted into a groove beneath head**
MOUTHPARTS: Chewing
EYES: Compound
WINGS: **Covered with scale or fine hairs; convex**
LEGS: Walking; tarsal formula 5-5-5
MISCELLANEOUS: **Head concealed from above**
HABITAT:
Adults: Flowers; decomposing animal materials
Larvae: **Decomposing plant and animal materials**
FOOD:
Adults: Decomposing matter; pollen
Larvae: **Decomposing matter; dried animal and plant materials; stored grains**

Adult dermestid

Dermestid larvae

Adult dermestid

Dermestids are scavengers, preferring decomposing materials that are high in protein. Many species cause economic damage. The larvae are generally recognized by their long body hairs and are at the development stage that causes the most damage.

The carpet beetle, *Anthrenus scrophulariae,* and the black carpet beetle, *Attagenus megatoma,* eat carpet, woolen materials, feathers, and other dried animal products and therefore are of concern to museums and owners of preserved animals. Most insect enthusiasts are aware of this destructive group of beetles and must take frequent measures to prevent them from destroying their prized bug collection.

Although dermestids are generally considered pests, one species, the hide beetle, *Dermestes maculatus,* is sometimes used by museum curators to thoroughly clean animal skeletons.

Soldier Beetles

Family: Cantharidae
(kan-THAIR-ih-dee)

Soldier beetle

Characteristics:

SIZE: $\frac{1}{8}$ to $\frac{5}{8}$ in. (3.2 to 15.9 mm)

SHAPE: **Elongated and parallel sided**

COLOR: Usually brown; with red, orange, or yellow markings on pronotum and wings

ANTENNAE: Threadlike; 11-segmented

MOUTHPARTS: Chewing

EYES: Compound

WINGS: **Soft and leatherlike**

LEGS: Walking; tarsal formula 5-5-5

MISCELLANEOUS: **Similar to fireflies except head extends beyond prothorax and no light-producing organs;** adults diurnal; larvae nocturnal

HABITAT:
- Adults: Typically found on flowers
- Larvae: Under bark and debris

FOOD:
- Adults: Predaceous on other small arthropods; pollen; nectar
- Larvae: Predaceous on other small arthropods

Soldier beetle

Soldier beetles resemble fireflies, but their head is not covered by the pronotum, and they do not possess a light-producing organ. These insects are common visitors to flowers, where they eat pollen, nectar, and/or other insect visitors.

The larvae are predaceous on soft-bodied insects such as caterpillars, maggots, and insect eggs and are generally more active at night.

Tiger Beetles

Family: Cicindelidae
(sih-sihn-DEHL-ih-dee)

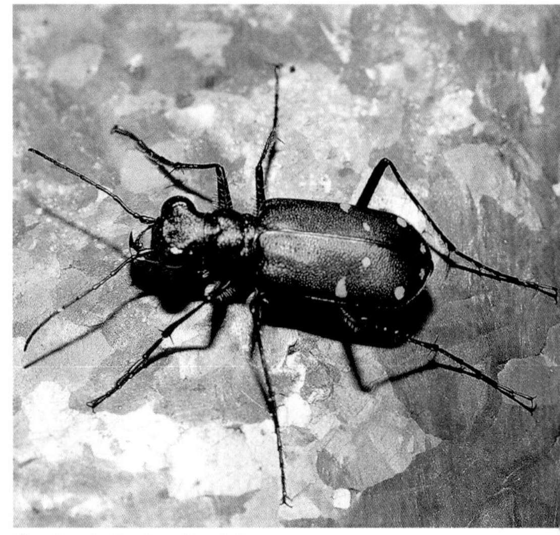
Tiger beetle (by Gary Brooks)

Characteristics:

SIZE: $\frac{1}{4}$ to $\frac{7}{8}$ in. (6.4 to 22.4 mm)
SHAPE: **Cylindrical; pronotum narrower than eyes and base of wings**
COLOR: **Often brightly colored; wings often with distinct color patterns**
ANTENNAE: Threadlike; inserted above base and in front of mandibles; 11-segmented
MOUTHPARTS: Chewing; sickle shaped
EYES: **Compound; large and bulging**
WINGS: **Forewings widen toward rear;** strong fliers
LEGS: Walking; l**ong, slender, with stout spines;** tarsal formula 5-5-5
MISCELLANEOUS: **Rapid runners;** adults mostly diurnal
HABITAT:
Adults: Terrestrial; **prefer sunny open areas**
Larvae: Vertical tunnels in soil; usually near water
FOOD:
Adults and Larvae: Predaceous on other arthropods

Tiger beetles are among the most beautiful and elusive insects. The adults are often iridescent blue, green, red, or a combination of these colors. Their beauty is matched by their speed both running and flying. Upon spotting a potential predator, they will dart a few feet away and stop in an almost mocking fashion. With its "catch me if you can" audacity, these agile insects can make a fool of most seasoned collectors.

Both the adults and larvae prey on a variety of smaller organisms. The adults use their speed to capture the prey, while the larvae hide in slender tunnels in open areas, waiting for a passing victim. The larva possesses a hump affixed with hooks at its middle back used to anchor itself to the tunnel.

Tiger beetle

Tiger beetle (by Gary Brooks)

Tumbling Flower Beetles

Family: Mordellidae
(mor-DEHL-ih-dee)

Characteristics:

SIZE: $\frac{1}{8}$ to $\frac{3}{8}$ in. (3.2 to 9.6 mm)
SHAPE: **Wedge shaped; humpbacked**
COLOR: **Black, brown, or gray; sometimes with light markings**
ANTENNAE: Threadlike, saw-toothed, or clubbed; 11-segmented
MOUTHPARTS: Chewing
EYES: Compound
WINGS: **Shorter than abdomen**
LEGS: Walking; femora often enlarged; tarsal formula 5-5-4
MISCELLANEOUS: **Abdomen pointed; head bent downward**
HABITAT:
Adults: **Flowers**
Larvae: Dead trees
FOOD:
Adults: Plant feeders
Larvae: Plant feeders; decomposing materials; predaceous on small arthropods

Tumbling flower beetles

Tumbling flower beetle

Tumbling flower beetle

Tumbling flower beetles are small, wedge-shaped beetles commonly found in flowers, particularly those in the sunflower group. When disturbed, the adults will tumble off the flower. Although the sight is quite amusing, it is also very frustrating for the avid collector trying to catch them.

The larvae are capable of boring into soft plant tissues and are found in decaying wood, the pith of some plants, or in bracket fungi. A few are leaf miners, some are predaceous, but most are believed to feed on decaying plant materials and the pith of weeds.

Twig Borers

Family: Bostrichidae
(bah-STRIHK-ih-dee)

Twig borer

Characteristics:

SIZE: $\frac{1}{8}$ to $\frac{1}{2}$ in. (3.2 to 12.8 mm)
SHAPE: **Elongated and somewhat cylindrical**
COLOR: **Generally dark**
ANTENNAE: Clubbed; 8- to 10-segmented, 3- to 4-segmented club
MOUTHPARTS: Chewing
EYES: Compound
Wings: Rounded at end
LEGS: Walking; tarsal formula 5-5-5
MISCELLANEOUS: **Head bent down, barely visible from above; pronotum large and rasplike**
HABITAT:
Adults: Living and dead trees
Larvae: Twigs
FOOD:
Adults and Larvae: Wood

Adult twig borers have a very distinctive shape. The head is tilted downward and not visible from above. The prothorax is large and possesses tubercles, giving it a rasplike appearance.

Twig borer

Adult twig borers bore into hardwood to create galleries in which to lay eggs. They often die at the entrance, thus preventing predators or parasites from entering. The hairless, C-shaped larvae feed on the wood, which may be living trees, dead twigs, or seasoned wood.

In the western United States, the lead cable borer, *Scobicia declivis,* normally feeds on oak, maple, or other tree species. However, on occasion, it may bore into the lead sheathing of telephone cables; it makes a one-sixteenth-inch hole that allows water in, causing short-circuiting. The lesser grain borer, *Rhyzopertha dominica,* can be a serious pest of stored grain.

Water Scavenger Beetles

Family: Hydrophilidae
(hy-droh-FIHL-ih-dee)

Water scavenger beetle antennae and palps

Characteristics:

SIZE: $\frac{1}{8}$ to $1\frac{1}{2}$ in. (3.2 to 38.1 mm)
SHAPE: **Oval**
COLOR: **Usually shiny black; some with yellowish margins on wings**
ANTENNAE: Short; clubbed; concealed; 4-segmented
MOUTHPARTS: Chewing; **maxillary palps longer than antennae**
EYES: Compound
LEGS: Adapted for swimming; flattened and fringed with hair; tarsal formula 5-5-5
MISCELLANEOUS: **Often possess a long keel ending in a sharp spine on the underside of the body**
HABITAT:
Adults and Larvae: Standing or running water; a few species in moist habitats
FOOD:
Adults: Usually decomposing materials; some predaceous on small arthropods
Larvae: Predaceous on small arthropods

Water scavenger beetle larva (by Gary Brooks)

Adult water scavenger beetle

Water scavenger beetles are similar to predaceous diving beetles but can be identified by physical and behavioral traits. Water scavenger beetles have long maxillary palps, which can be confused with their small and typically concealed antennae.

These beetles rise to the water surface head first to breathe and use their antennae to transfer fresh air to fine hairs on their underside, resulting in a silvery appearance. This air is used for extended underwater trips.

Most species live in shallow, weedy margins of stagnant water. A few, however, live and breed in dung or decaying vegetable matter. The adults feed mainly on decomposing material, while the larvae are voracious predators.

Weevils, Snout Beetles

Family: Curculionidae
(kur-kyoo-lee-AHN-ih-dee)

Characteristics:

SIZE: $\frac{1}{8}$ to 1 in. (3.2 to 25.4 mm)
SHAPE: Oval or cylindrical
COLOR: Variable
ANTENNAE: **Elbowed and clubbed; inserted into the snout**
MOUTHPARTS: Chewing; **located on the end of the snout**
EYES: Compound
WINGS: Typically well developed
LEGS: Walking; tarsal formula apparently 4-4-4, actually 5-5-5
HABITAT:
- Adults: Plants
- Larvae: Plants; typically inside tissue

FOOD:
- Adults: Plant roots; stems; flowers; fruit or seeds
- Larvae: Internal plant tissues

Boll weevil pupa

Acorn weevil

Goldenheaded weevil, *Compsus auricephalus*

Adult boll weevil, *Anthonomus grandis*

The weevil family has over forty thousand species worldwide, making it the largest family of living organisms. These insects are generally easy to recognize by the prolonged snout and elbowed antennae.

All weevils are plant feeders. Many are serious pests, including the cotton boll weevil, *Anthonomus grandis;* the pecan weevil, *Curculio caryae;* the alfalfa weevil, *Hypera postica;* and the plum curculio, *Conotrachelus nenuphar.*

Eggs are normally laid inside plant tissue, and the C-shaped larva completes development in this protected environment.

Whirligig Beetles

Family: Gyrinidae
(jy-RIHN-ih-dee)

Characteristics:

SIZE: $\frac{5}{16}$ to $\frac{1}{2}$ in. (8.0 to 12.8 mm)
SHAPE: **Body flattened; oval**
COLOR: **Generally black**
ANTENNAE: Short; clubbed; 8-segmented
MOUTHPARTS: Chewing
EYES: Compound; **divided at the water line**
WINGS: Present; good fliers
LEGS: **Modified for swimming; forelegs long; middle and hind legs short and paddlelike;** tarsal formula 5-5-5
MISCELLANEOUS: **Usually in groups on surface of quiet water**
HABITAT:
Adults: Aquatic; shallow margins
Larvae: Aquatic
FOOD:
Adults: Predaceous on other small arthropods
Larvae: Predaceous on other underwater organisms

Whirligig beetles are gregarious, aquatic insects that often form large swarms on the water surface. They are rapid swimmers and typically swim in endless zigzag patterns. These beetles are unique because their eyes are divided, giving them the ability to observe both below and above the water surface.

Adult whirligig beetles are common in ponds and quiet streams. They are either predaceous or scavengers on surface insects; the larvae are predaceous on underwater insects and can be cannibalistic when other prey is limited.

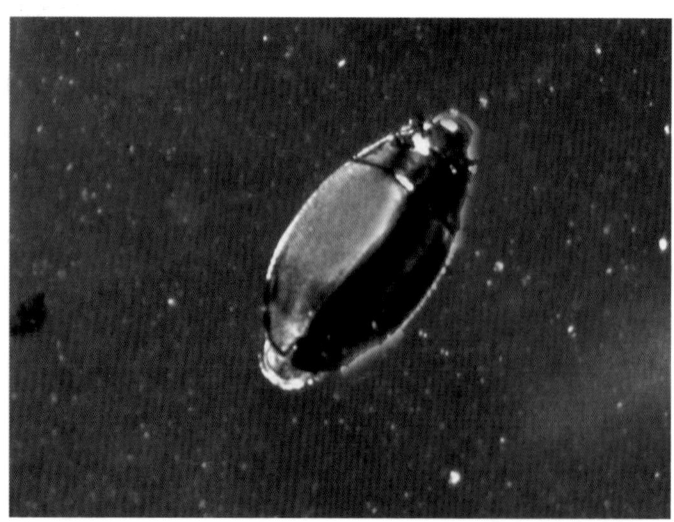

Whirligig beetle

Scorpionflies

Order: Mecoptera
(mee-KAHP-tur-uh)
meco, long; *ptera,* wing

Characteristics:

SIZE: ½ to 1 in. (12.8 to 25.4 mm)

SHAPE: Slender; **head elongated and beaklike**

COLOR: **Reddish; spotted with black markings**

ANTENNAE: Threadlike

MOUTHPARTS: Chewing

EYES: Compound; present or absent

WINGS: Membranous; many cross veins; forewings and hind wings similar size

LEGS: Walking

MISCELLANEOUS: **Male genitalia of some species enlarged and may resemble a scorpion stinger**

HABITAT:
- Adults: **Trees**
- Larvae: Soil litter

FOOD:
- Adults: Dead insects; predaceous on small arthropods; nectar
- Larvae: Dead insects; predaceous on small arthropods

Scorpionfly

Scorpionfly

METAMORPHOSIS: Complete (egg, larva, pupa, adult)

Scorpionflies can be frightening insects but are actually quite harmless. The adults are characterized by four wings that are the same size and shape; long, threadlike antennae; and chewing mouthparts at the end of an elongated beak. Their common name is derived from the swollen male genitalia of some species, which are held curved over their back in scorpion fashion. These insects neither sting nor bite.

The female scorpionfly lays her eggs among ground litter. The larvae resemble spiny caterpillars, having three pairs of true legs and four to ten pairs of prolegs. These immature insects live in leaf litter or under rocks and feed on decomposing insect parts.

Fleas

Order: Siphonaptera
(sy-fuh-NAHP-tur-uh)
siphon, tube; *aptera,* without a wing

Characteristics:
SIZE: $\frac{1}{16}$ to $\frac{1}{8}$ in. (1.6 to 3.2 mm)
SHAPE: **Laterally compressed; oval**
COLOR: Dark
ANTENNAE: Short
MOUTHPARTS: Sucking
EYES: Compound; present or absent
WINGS: **Absent**
LEGS: Relatively long; coxae large
MISCELLANEOUS: **Jumping insects; many backward-projecting bristles**
HABITAT:
Adults: **Mammals;** few feed on birds
Larvae: **Bedding of adult host animal**
FOOD:
Adults: Blood
Larvae: Organic materials
METAMORPHOSIS: Complete (egg, larva, pupa, adult)

Anyone who has had a cat or dog has probably had an annoying experience with fleas. These wingless insects are laterally compressed and have long, spiny jumping legs.

Fleas are ectoparasites of mammals. The female flea lays her eggs while feeding on the host; the eggs drop onto the bedding of the host. The larvae feed on the waste products of the adult flea or on pet dander. After maturing, the larva spins a silken cocoon and pupates. The pupa can lie dormant for extended periods and may not develop into the adult stage for seven days up to one year.

There are several species of fleas in Texas, including the cat flea, *Ctenocephalides felis;* the dog flea, *C. canis;* and the oriental flea, *Xenpsylla cheopis.* The cat flea is most common and will readily bite dogs and humans.

Historically, the flea has been an important vector of such diseases as bubonic plague and typhus. Fleas also serve as intermediate hosts for tapeworms, which infect dogs and occasionally humans.

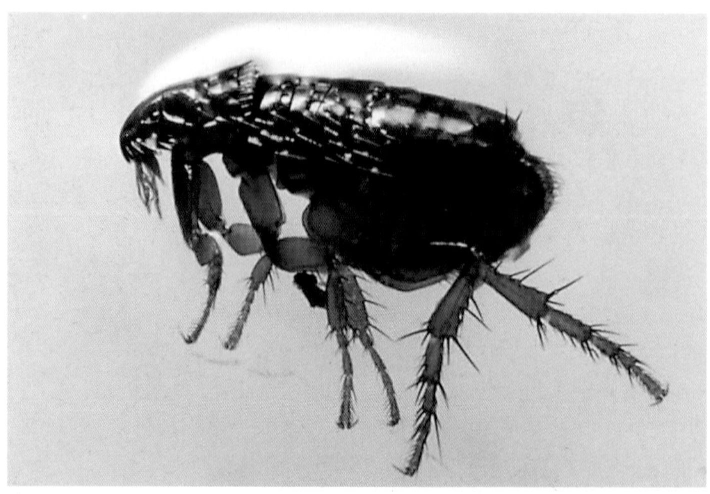

Flea

Adults: Flies

Larvae: Maggots

Order: Diptera (DIHP-tur-uh)

di, two; *ptera,* wing

Characteristics:

SIZE: Tiny to ~1 in. (2.5 cm)
ANTENNAE: Variable
MOUTHPARTS: Various adaptations for sucking
EYES: Compound; often large
WINGS: **One pair; located on mesothorax; metathoracic wings reduced and knoblike**
LEGS: Typically for walking
HABITAT: Extremely variable
FOOD: Predators or parasitoids on other arthropods; parasites; scavengers; a few plant feeders
METAMORPHOSIS: Complete (egg, larva, pupa, adult)

The Diptera are a large and diverse group differing from most other groups of insects by having only one pair of wings, which are located on the mesothorax. The second pair of wings is reduced to knoblike structures, called halteres, that are believed to serve the insect as balancers.

The larvae, commonly called "maggots," are generally legless and wedge shaped; they typically have no distinct head (mosquito larvae are a notable exception). The mouthparts of maggots are at the tapered end of the body, while the breathing apparatus (spiracles) are located on the broad end. Many species live and feed in various forms of decomposing materials.

The Diptera are a very important order of insects. Many species are parasitic and feed on humans and animals. Some of these species, particularly mosquitoes, are vectors of many of the world's most devastating diseases. Other species are parasitoids of other insects and are important in keeping these insect numbers in check.

COMMON FAMILIES OF THE ORDER DIPTERA, WITH THEIR PREFERRED HABITATS AND FEEDING CHARACTERISTICS

Family	*Near water*	*Decomposing materials*	*Plants*	*Animals*	*Predators or parasitoids*	*Page*
Bee flies: Bombyliidae			A		L	139
Blow flies: Calliphoridae		L, A		L		140
Crane flies: Tipulidae	A	L				141
Flesh flies: Sarcophagidae		L, A	A		L	142
Frit flies, eye gnats: Chloropidae		L, A	L	A		143
Horse flies, deer flies: Tabanidae	L, A			A		144
House flies, stable flies, horn flies: Muscidae		L, A		A		145
Leaf miner flies: Agromyzidae			L, A			146
Long-legged flies: Dolichopodidae			L, A			147
Louse flies: Hippoboscidae				L, A		148
March flies: Bibionidae		L	A			149
Midges: Chironomidae	L, A		L	A		150
Mosquitoes: Culicidae	L			A		151
Moths, sand flies: Psychodidae	A	L, A		A		152
Parasitic flies: Tachinidae			A		L	153
Robber flies: Asilidae		L			A	154
Soldier flies: Stratiomyidae	L	L, A	A			155
Syrphid or flower flies: Syrphidae		L	A		L	156

Note: L: larvae; A: adults.

Bee Flies

Family: Bombyliidae
(bahm-bih-LEE-ih-dee)

Characteristics:

SIZE: $\frac{1}{4}$ to $\frac{3}{8}$ in. (6.4 to 9.6 mm)
SHAPE: **Often resemble bees; stout bodied and very hairy**
COLOR: Variable
ANTENNAE: Short
MOUTHPARTS: **Sometimes elongated and beaklike**
EYES: Compound
WINGS: Clear, **often with patterns; typically held outstretched when at rest**
LEGS: Walking
MISCELLANEOUS: **Often hover near flowers**
HABITAT:
- Adults: **Flowers**
- Larvae: Soil; inside or on host

FOOD:
- Adults: Nectar; pollen
- Larvae: External or internal parasitoids of soil-inhabiting insects; predaceous on grasshopper eggs

Bee fly

These fuzzy, beelike insects are often observed hovering over flowers, where the adults feed on nectar and pollen. When disturbed, they will dart several feet away and continue to hover. Although they mimic bees, bee flies neither sting nor bite.

Larvae of most species are parasitoids on the soil-inhabiting immature stages of Hymenoptera, Lepidoptera, Coleoptera, Neuroptera, and Diptera. A few species, however, are predaceous on short-horned grasshopper eggs (order Orthoptera, family Acrididae).

Bee fly

Bee fly

Blow Flies, Bottle Flies

Family: Calliphoridae
(kal-ih-FOR-ih-dee)

Characteristics:
SIZE: ~$\frac{3}{8}$ in. (9.6 mm)
SHAPE: **Similar to that of house flies**
COLOR: **Often metallic blue, green, or bronze**
ANTENNAE: Aristate
MOUTHPARTS: Sucking
EYES: Compound
WINGS: Clear
LEGS: Walking
HABITAT:
Adults: **Dung; decomposing materials; flowers**
Larvae: Decomposing materials
FOOD:
Adults: Nectar; pollen; decaying materials
Larvae: Scavengers, mostly of decaying flesh

Blow flies and bottle flies are strong fliers and attracted to freshly dead animals or fresh manure. Forensic entomologists can often accurately determine the time of death by identifying the species and stage of development of calliphorid flies that are present on a corpse.

Blow fly

Blow fly

Blow fly

Blow fly (by Gary Brooks)

Blow flies seldom enter homes, so their presence inside a house may indicate that a small varmint may have died somewhere nearby and needs disposal. Carl Linnaeus, the great taxonomist, showed his wonderful sense of humor (or vivid imagination) when he named the common blue bottle fly *Calliphora vomitoria.*

Most species of calliphorids feed on dead tissue. However, one species of note, *Cochliomyia hominivorax,* is known as the "primary screwworm." The larvae of this species feed on living tissue and have caused tremendous economic loss to the livestock industry in the southern United States. After World War II, scientists discovered that radiation could be used to sterilize male screwworms, which would then compete with native males for females. The females that mated with the sterile males produced nonviable eggs. As a result of this work, the screwworm has been virtually eradicated from the United States and much of Mexico.

Adults: Crane Flies
Larvae: Leatherjackets
Family: Tipulidae
(ty-PYOO-lih-dee)

Characteristics:
SIZE: $\frac{1}{8}$ to >1 in. (3.2 to 25.4 mm)
SHAPE: **Resemble large mosquitoes; humpbacked thorax**
COLOR: **Brownish or gray**
ANTENNAE: Threadlike
MOUTHPARTS: Prominent
EYES: Compound
WINGS: One pair; long and slender; may have dark patterns; **prominent halteres**
LEGS: Walking; **very long and fragile**
MISCELLANEOUS: **V-shaped suture on top of thorax**
HABITAT:
Adults: Near water or thick vegetation
Larvae: Aquatic or moist soil
FOOD:
Adults: Generally do not feed
Larvae: Decaying material; occasionally predaceous on other small arthropods

Crane fly

Crane fly (note V-shaped suture)

Crane fly

Every spring, huge mosquito-like insects are attracted to lights around homes. These fierce-looking creatures are actually nonbiting crane flies, so named because of their long, fragile legs. The larvae, sometimes referred to as "leatherjackets" due to their thick skin, live in moist, poorly drained soil that has abundant organic matter. Occasionally, the larvae feed on the roots of crop plants and may cause damage.

The adults are weak fliers and attracted to lights. They do not bite. In fact, some feed on plant nectar for a short time; other species do not feed. Most adult crane flies live only a few days.

Flesh Flies

Family: Sarcophagidae
(sar-koh-FAJ-ih-dee)

Characteristics:

SIZE: $\frac{1}{4}$ to $\frac{1}{2}$ in. (6.4 to 12.8 mm)
SHAPE: **Similar to that of house flies**
COLOR: Never metallic; **often black with stripes;** sometimes with checkerboard pattern on abdomen; **often with red on tip of abdomen**
ANTENNAE: Aristate; arista feathery on bottom half
MOUTHPARTS: Sucking
EYES: Compound
WINGS: Transparent; without spots
LEGS: Walking
HABITAT:
Adults: **Flowers**
Larvae: Animal flesh; excrement (particularly dog stools)
FOOD:
Adults: Nectar; honeydew
Larvae: Dead or living animal flesh; a few rob wasp nests, destroy eggs, and feed on the provisions of the nest; parasitoids on insects

Flesh fly

Flesh fly

This is a rather diverse group of flies. Although referred to as "flesh flies," many species do not feed on animal tissues. Some species are parasitoids on other insects, while others are parasites of vertebrate species.

The territorial males will land on tall plants and wait to mate with a passing female.

Unlike eggs of most insects, the eggs of sarcophagid flies hatch inside the female; she will subsequently give birth to living first-instar larvae. This unique process may provide an advantage to the young flesh fly larvae. Blow flies (order Diptera, family Caliphoridae) typically are the first to locate and lay eggs on a fresh animal carcass. As blow fly larvae hatch, space becomes limited on the carcass, which may prevent late arrivers from finding a suitable location to feed. However, since the female flesh fly will deposit her brood as young larvae, they are ready to compete for the very limited space.

Frit Flies, Eye Gnats

Family: Chloropidae
(klor-AHP-ih-dee)

Characteristics:
SIZE: **$\frac{1}{16}$ to $\frac{1}{8}$ in. (1.6 to 3.2 mm)**
SHAPE: Typically flylike
COLOR: Variable; **often colorful; black or black and yellow**
ANTENNAE: Aristate
MOUTHPARTS: Sponging
EYES: Compound
WINGS: Clear
LEGS: Walking
MISCELLANEOUS: Usually bare
HABITAT:
Adults: **Grassy areas**
Larvae: Grass stems; decaying vegetable and animal matter
FOOD:
Adults: **Females, exudates from open sores; males, plant juices**
Larvae: Decaying vegetable matter; grass stems; some predaceous on small arthropods

Gnats on dog's eye

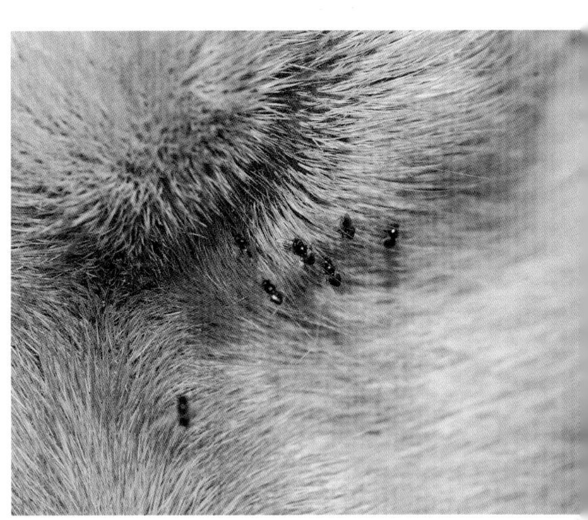

Gnats on dog's eye

Eye gnats, *Hippelates* spp., are tiny, very annoying insects that are sometimes called "grass flies" for their tendency to be in open, grassy areas. Female eye gnats are attracted to eyes and open sores, where they feed on mucus, pus, and other secretions. In addition to being annoying, they can transmit diseases, including pink eye, anaplasmosis, and bovine mastitis.

The adult females lay eggs in moist soil, where the larvae feed on decaying plant material.

Horse Flies, Deer Flies

Family: Tabanidae
(tuh-BAN-ih-dee)

Horse fly

Characteristics:

SIZE: ½ to >1 in. (12.8 to 25.4 mm)
SHAPE: **Stout bodied; large head**
COLOR: **Generally dark; sometimes gray, striped, or spotted**
ANTENNAE: Swollen at base and taper to a point
MOUTHPARTS: **Short, thick, and sword shaped**
EYES: **Compound eyes, very pronounced; cover most of the head; often iridescent**
WINGS: Large; sometimes spotted
LEGS: Walking
MISCELLANEOUS: Usually bare
HABITAT:
Adults: **Swamps and low-lying areas**
Larvae: Aquatic or in moist soil
FOOD:
Adults: **Females, blood and nectar; males, nectar**
Larvae: Predaceous on other arthropods

Horse flies are among the largest flies. Some species are over an inch long (2.54 cm) and have a wing span of two inches (5.2 cm). They have large heads that are covered almost entirely by their compound eyes. The eyes meet in the middle of the male's head but are separated in females. The eyes of many species have a zigzag iridescent pattern.

Tabanids prefer tree- or brush-covered areas near creeks and streams, where the males feed on nectar. Female horse flies can inflict a painful bite, slashing the skin and sucking the blood, which is necessary for proper egg production. In addition, these fast fliers can mechanically transmit the pathogens that cause anthrax, equine infectious anemia, and other diseases.

Horse fly and deer fly larvae generally live in saturated soils or in aquatic habitats. They feed on small invertebrates and, occasionally, even small vertebrates such as toads.

The two common genera in Texas are *Tabanus* spp., the horse flies, which are typically large and black; and *Chrysops* spp., the deer flies, which are smaller and generally lighter colored.

Horse fly

Deer fly

House Flies, Stable Flies, Horn Flies

Family: Muscidae
(MUHS-kih-dee)

Stable fly, *Stomoxys calcitrans*

Characteristics:

SIZE: $\frac{1}{10}$ to $\frac{1}{3}$ in. (2.5 to 8.5 mm)
SHAPE: **Typically the shape of house flies**
COLOR: **Grayish; dark; never metallic**
ANTENNAE: Aristate; arista feathery the entire length
MOUTHPARTS: Variable
EYES: Compound
WINGS: Clear
LEGS: Walking
HABITAT:
Adults: Variable
Larvae: **Excrement; decaying material**
FOOD:
Adults: Scavengers; blood
Larvae: Excrement; decaying material; predaceous on small arthropods

There are over six hundred species of muscids in North America, many of which are not only a nuisance but also potential disease vectors. Among the more infamous species are *Musca domestica,* house flies; *M. autumnalis,* face flies; *Stomoxys calcitrans,* stable flies; and *Haematobia irritans,* horn flies. Most species are prolific breeders and can reach enormous numbers if conditions are ideal.

Each species has its own particular niche, and knowledge of their ecology is important in the proper management of these pests. For example, horn flies prefer open pastures and breed only in freshly deposited manure. Stable flies, however, prefer animals in confinement and breed in a mixture of manure and straw or feed. Sanitation is often a key management component.

House flies, *Musca domestica*

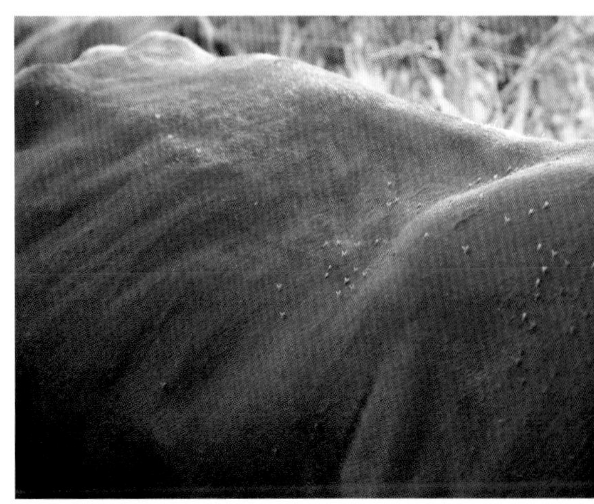

Horn flies, *Haematobia irritans*

Leaf Miner Flies

Family: Agromyzidae
(ag-roh-MY-zih-dee)

Leaf miner fly tunnels

Characteristics:
SIZE: **<$\frac{1}{8}$ in. (3.2 mm)**
SHAPE: Flylike
COLOR: Black or gray, often with yellow
ANTENNAE: Aristate
MOUTHPARTS: Sucking
EYES: Compound
WINGS: Transparent
LEGS: Walking
MISCELLANEOUS: **Often can be identified by their mines in leaves**
HABITAT:
Adults: Plants
Larvae: **Plants; leaves; a few in plant stems**
FOOD:
Adults: Plant fluids
Larvae: Plant feeders

These tiny flies are better known by the plant damage made by the larvae rather than by the appearance of the adult fly. Both woody and herbaceous plants can serve as hosts, although most species are host selective.

Female leaf miners lay their eggs in plant tissue, where the larvae will feed between the epidermal layers. The resulting mines typically are winding and enlarge as the larvae grow. Some species pupate within the mine, while others tunnel out of the leaf and drop to the ground. Though sometimes unsightly, the damage caused by these insects is generally not harmful to the plant. Other families of Diptera, as well as members of the orders Lepidoptera, Coleoptera, and Hymenoptera, also have plant-mining species.

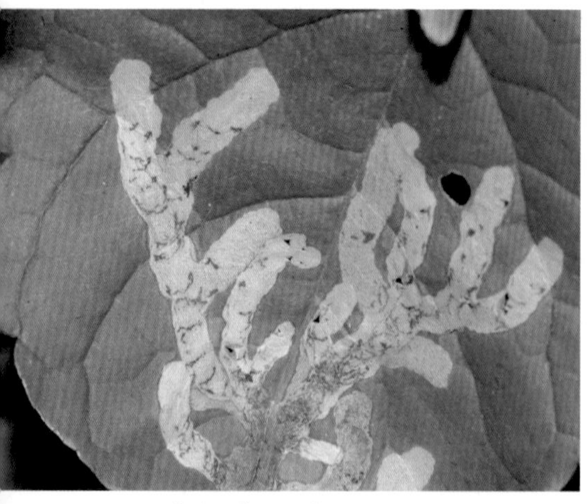

Leaf miner fly tunnels

Long-legged Flies

Family: Dolichopodidae
(duh-lihk-oh-POH-dih-dee)

Characteristics:

SIZE: ~$\frac{1}{8}$ in. (3.2 mm)
SHAPE: Typically flylike but more slender and tapering
COLOR: **Commonly metallic green or blue**
ANTENNAE: Aristate
MOUTHPARTS: Short
EYES: Compound
WINGS: Translucent; sometimes spotted
LEGS: Walking; **long; slender**
MISCELLANEOUS: **Common in shaded areas; males often have large genitalia**
HABITAT:
 Adults: Meadows and marshes
 Larvae: Wet soil; under bark
FOOD:
 Adults and Larvae: Predaceous on small arthropods

Long-legged fly

These small, colorful flies are commonly seen on foliage and, as the name implies, have legs that are relatively long for their bodies. Long-legged flies are generally metallic and often quite colorful. The males attempt to attract females through a ritual dance but sometimes, perhaps out of frustration, will mate with any object of similar size.

Both adults and larvae are predaceous on small insects. The larvae of some species live under the bark of trees and are important predators of bark beetle larvae.

Long-legged fly

Long-legged fly

Louse Flies, Keds

Family: Hippoboscidae
(hihp-oh-BAHS-kih-dee)

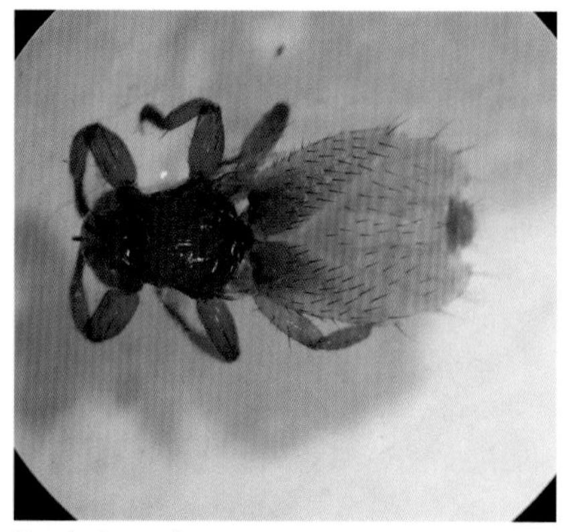
Deer ked, top view

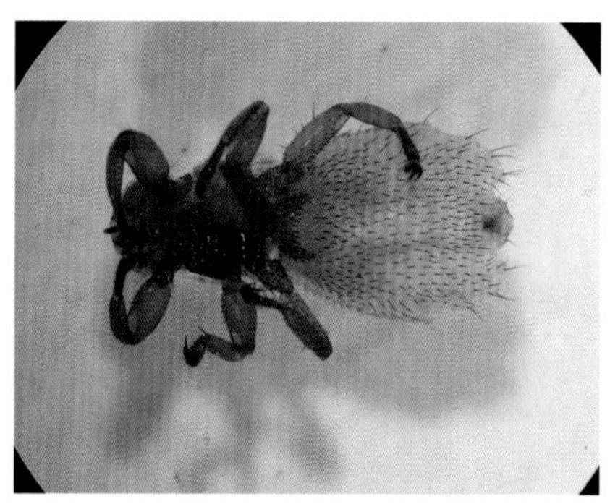
Deer ked, bottom view

Characteristics:
SIZE: ~$\frac{1}{4}$ in. (6.4 mm)
SHAPE: **Flattened; ticklike**
COLOR: Typically dark brownish
ANTENNAE: **Sunken; indistinguishable**
MOUTHPARTS: **Typically projecting forward**
EYES: Compound
WINGS: **Wings usually broken off (parasites of birds); wingless (parasites of mammals)**
LEGS: Walking
HABITAT:
Adults: **Ectoparasites**
Larvae: Live in pouch of female until ready to pupate
FOOD:
Adults: Blood feeders mostly on birds and a few mammals
Larvae: Fed by female

Some species of louse flies have wings that are shed soon after finding a suitable host; other species are wingless. Both sexes are blood feeders, although they appear to cause little distress to the host. Most species are very host specific, but the winged species often can feed on more than one type of host.

The eggs hatch and the larvae develop inside the uterus of the female ked. When fully developed, the immobile, nonparasitic larvae are deposited on the host, where they do not feed but quickly form puparia.

Birds are the most common host, but other warm-blooded animals, particularly deer, also serve as hosts. The sheep ked, *Melophagus ovinus,* is a fairly common ectoparasite of sheep and goats but occasionally bites humans who handle these animals.

March Flies, Lovebugs

Family: Bibionidae
(bih-bee-AHN-ih-dee)

Characteristics:

SIZE: $\frac{1}{3}$ in. (8.5 mm)
SHAPE: **Stout bodied; hairy; large head**
COLOR: **Generally black or dark colored; often with yellow or red thorax**
ANTENNAE: Short; arise from low on face
MOUTHPARTS: Long and narrow in female
EYES: Compound
WINGS: **Dark;** clear
LEGS: Walking
MISCELLANEOUS: **Often in large numbers and flying joined as pairs**
HABITAT:
Adults: Flowers
Larvae: Decaying plant materials
FOOD:
Adults: Decomposing materials or pollen
Larvae: Decomposing plant material

March flies can be abundant anytime in the spring and early summer and typically live only a few days. The males are commonly observed flying in large swarms, and the females are attracted to the swirling mass, where they may find a suitable mate. The pair flies off in tandem, thus giving them another common name: "lovebugs." Mated females dig holes in moist soil and deposit two hundred to three hundred eggs. The larvae feed on decomposing materials and overwinter in this stage.

March flies mating (by Gary Brooks)

Large numbers of these insects are sometimes attracted to automobile exhaust. When they strike the vehicle windshield, the smashed female releases her eggs and may cause a large smear on the windshield, often obscuring the driver's vision.

March fly

Midges

Family: Chironomidae
(ky-roh-NAH-mih-dee)

Characteristics:

SIZE: $\frac{1}{10}$ to $\frac{1}{2}$ in. (2.4 to 12.8 mm)
SHAPE: **Mosquito-like**
COLOR: Variable
ANTENNAE: Female, threadlike; **male, feathery**
MOUTHPARTS: **Short; nonbiting**
EYES: Compound
WINGS: **Long and narrow; without scale**
LEGS: Long
MISCELLANEOUS: **Usually hold their forelegs up when at rest;** adults attracted to lights

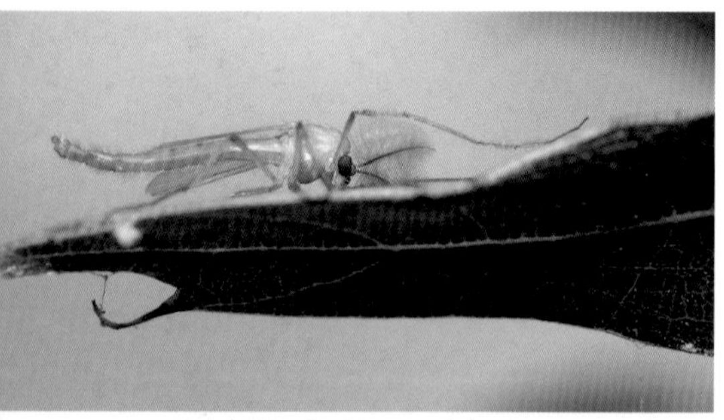

Male midge (by Gary Brooks)

Female midge

HABITAT:
Adults: **Often in large swarms near water**
Larvae: Aquatic; inhabit mud at bottom of lakes and rivers

FOOD:
Adults: Some species, organic matter; others, nonfeeding
Larvae: Microscopic animals and plants; organic matter

Have you ever been at the lake on a beautiful spring night, walked by a lantern, and had what seem to be thousands of tiny bugs fly into your ears and up your nose? These nonbiting, but very often aggravating, insects are midges. Often mistaken for mosquitoes, midges can be easily identified by the males' fuzzy antennae and by the way they hold their forelegs up and out when at rest.

Midges commonly occur in large swarms near water, such as lakes and rivers, and are attracted to any light source. When swarming, they produce a humming sound that can be heard for considerable distances.

The larvae, which live in the muddy bottom of aquatic habitats, also occur in large numbers and are an important food source for many small fish and other aquatic organisms. Members of the genus *Chironomus* are particularly interesting because the immature stage is bright red as a result of hemoglobin in their blood. These tiny wormlike larvae, called "bloodworms," are often sold as food for pet fish.

Mosquitoes

Family: Culicidae
(kyoo-LIHS-ih-dee)

Characteristics:

SIZE: $\frac{1}{8}$ to $\frac{3}{8}$ in. (3.2 to 9.6 mm)
SHAPE: Typically mosquito-like
COLOR: **Variable**
ANTENNAE: Female, threadlike with short circlets of hair; male, feathery
MOUTHPARTS: **Females, piercing/sucking and long; male, sucking**
EYES: Compound
WINGS: **Narrow; scales along veins and wing margins**
LEGS: Walking
HABITAT:
Adults: Associated with water
Larvae: **Aquatic**
FOOD:
Adults: **Females, blood and nectar; males, nectar**
Larvae: Mostly organic debris; a few predaceous

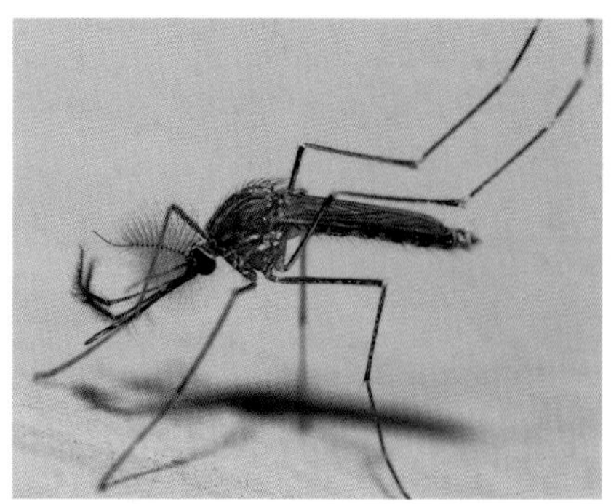
Male mosquito

No other organism in history has been associated with so much suffering and death as mosquitoes. Over 150 species occur in the United States of which more than 50 reside in Texas. In addition to be being a nuisance, many of these species can transmit deadly pathogens that cause diseases such as malaria, yellow fever, and West Nile virus. Only the female mosquito feeds on blood, which is necessary for egg development. The males cannot bite and feed on nectar.

Mosquitoes must have water to develop, but the type of habitat varies. The southern house mosquito, *Culex pipens quinquefasciatus,* for example, is common in Texas and develops in standing, polluted water. The Asian tiger mosquito, *Aedes albopictus,* however, is a recent immigrant to the United States and breeds in old tires, cans, and other small places that can hold water.

Mosquito larvae are called "wrigglers" because of their twisting motion when they are disturbed. The pupae, called "tumblers," are not as active and roll when they are disturbed. Mosquito larvae feed mainly on organic matter and microorganisms, but members of the genus *Toxorhynchites* are predaceous as larvae and often feed on other mosquito larvae. The adult females of this genus feed on nectar rather than blood and can be recognized by their proboscis, which curves sharply downward.

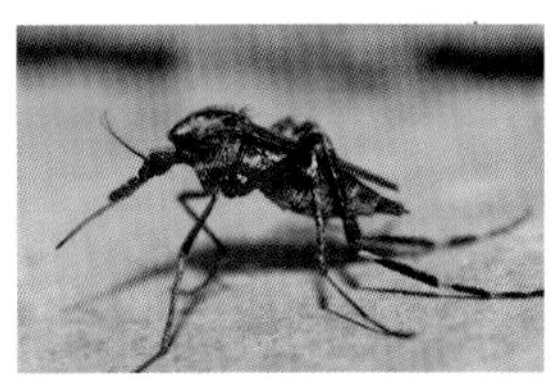
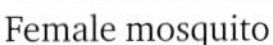
Female mosquito

Mosquito larvae

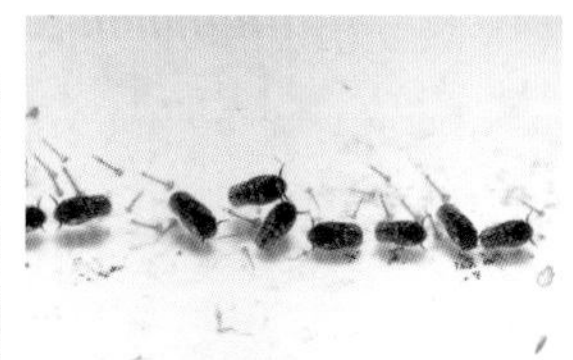
Mosquito pupae

Moth Flies

Family: Psychodidae
(sy-KOH-dih-dee)

Moth fly

Characteristics:

SIZE: ~$\frac{1}{6}$ in. (4.2 mm)
SHAPE: **Resemble small moths**
COLOR: Tan to dark gray
ANTENNAE: **Feathery**
EYES: Compound
WINGS: One pair; **hairy; broad and pointed; held rooflike when at rest**
LEGS: Walking
MISCELLANEOUS: **Very hairy bodies**
HABITAT:
Adults: **Moist, shady areas; drains, sewers, plumbing fixtures**
Larvae: Decaying matter; water
FOOD:
Adults: Most do not feed; some decaying material; a few bloodsuckers
Larvae: Decaying matter; microscopic plants and animals

Moth fly

Sometimes known as "drain flies" or "sewer flies," moth flies are commonly observed in moist, dimly lit areas such as urinals, drains, and sewers. The female moth flies lay their eggs on the surface film of sewage water and other places with a high content of organic materials. The larvae feed on this material and sometimes can be seen emerging from drains.

Sand flies, *Phlebotomus* spp., occur in the tropics, subtropics, and some parts of the southern United States. They are vicious biters that can transmit several disease pathogens.

Parasitic Flies, Tachinid Flies

Family: Tachinidae
(tuh-KY-nih-dee)

Characteristics:

SIZE: $\frac{1}{4}$ to $\frac{2}{3}$ in. (6.4 to 16.9 mm)
SHAPE: Variable; most resemble large house flies; some resemble bees
COLOR: Typically gray; some metallic
ANTENNAE: Aristate; arista bare
MOUTHPARTS: Sucking
EYES: Compound
WINGS: Transparent
LEGS: Walking
MISCELLANEOUS: **Heavy body bristles, particularly at tip of abdomen**
HABITAT:
Adults: **Typically, plants**
Larvae: Insect hosts
FOOD:
Adults: Nectar; pollen
Larvae: Internal parasitoids of larvae of Coleoptera and Lepidoptera and nymphs of Hemiptera and Orthoptera; a few parasitoids of adult insects

Tachinid fly

Leaf-footed bug with tachinid egg

Tachinid fly

Tachinid flies are large, typically hairy, and very common flies. These flies can often be separated from others by the presence of stout hairs on the tip of the abdomen.

Tachinid flies are important because most species are internal parasitoids on the immature stages of other insects. The adult female will generally lay her eggs directly on the surface of the host, and the newly hatched larvae will bore inside. Other species inject their eggs directly into the host. Still others lay their eggs on leaf surfaces that are subsequently eaten by the host.

Robber Flies

Family: Asilidae
(uh-SIHL-ih-dee)

Characteristics:

SIZE: $\frac{1}{4}$ to $1\frac{1}{4}$ in. (6.4 to 31.8 mm)
SHAPE: **Thorax relatively large; abdomen often slender; some resemble bees**
COLOR: **Variable**
ANTENNAE: **Stylus shaped**
MOUTHPARTS: **Pointed; stout; projecting downward from head**
EYES: Compound; large
WINGS: Translucent
LEGS: Long and bristled
MISCELLANEOUS: **Top of head hollowed between eyes; face bearded; generally hairy**
HABITAT:
Adults: **Open areas**
Larvae: Soil and decaying materials
FOOD:
Adults and Larvae: Predaceous on other arthropods

Bee mimicking robber fly, *Laphria* spp.

Robber fly

Robber flies typically fly close to the ground in a zigzag pattern as they search for prey. They can catch insects in flight and do not seem to be particularly choosy about what they catch, or even if the prey is larger than they are. Robber flies are particularly common in the southwestern United States and Texas.

Some members of the genus *Laphria* resemble honey bees but are distinguished in the field by their erratic flight pattern. It is not known if this mimicry is to protect the fly or to allow the robber fly to get close to its potential food source. Another genus, *Diogmites,* is known as the hanging thief because of its habit of hanging from plants by its long forelegs.

Robber fly beard

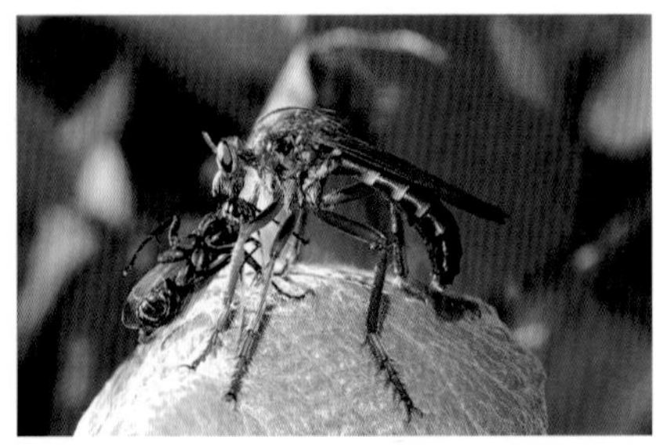

Robber fly eating prey (by Gary Brooks)

Soldier Flies

Family: Stratiomyidae
(stra-tee-oh-MY-ih-dee)

Soldier fly

Soldier fly

Characteristics:

SIZE: $\frac{1}{3}$ to $\frac{3}{4}$ in. (8.5 to 19.2 mm)
SHAPE: **Often broad and flattened; sometimes beelike or wasplike**
COLOR: **Most species dark; some with yellow, green, blue, or black striped abdomen; sometimes metallic**
ANTENNAE: **Often held in a Y-shaped position**
MOUTHPARTS: Sucking
EYES: Compound
WINGS: Folded over back when at rest
LEGS: Walking
MISCELLANEOUS: **Abdomen sometimes flat and broad, sometimes elongated; typically extends beyond the sides of the wings**
HABITAT:
Adults: **Flowers**
Larvae: Aquatic; decaying vegetable material; dung
FOOD:
Adults: Pollen; nectar; honeydew
Larvae: Algae; scavengers; predaceous on other arthropods

Adult soldier flies are weak fliers often found in low-growing vegetation and flowers, particularly in the carrot family. The larvae, depending upon species, can live in a variety of habitats. Some species can tolerate high salinity, while others prefer human or animal dung. A few species live under tree bark, feeding on microflora and fauna. The larvae are filter feeders, siphoning small pieces of organic matter and algae from the moist environment in which they live.

The black soldier fly, *Hermetia illucens,* is considered by most experts to be beneficial. This species lives in manure and other decaying organic matter, reducing the quality of this habitat as a breeding site for more pesky species such as the common house fly and stable flies. However, this species has been reported to cause human enteric myiasis, a condition in which the fly larvae live in human digestive systems.

Soldier fly pupa

Syrphid Flies, Flower Flies

Family: Syrphidae
(SUR-fih-dee)

Characteristics:

SIZE: $\frac{1}{4}$ to $\frac{1}{2}$ in. (6.4 to 12.8 mm)
SHAPE: Typically fly shaped
COLOR: **Generally black with yellow bands; often resemble bees or wasps**
ANTENNAE: Aristate
MOUTHPARTS: Sucking
EYES: Compound
WINGS: Spurious vein between radius and media
LEGS: Walking
MISCELLANEOUS: Many hairy, others hairless; **often seen hovering above flowers**
HABITAT:
Adults: **Typically flowers**
Larvae: Plants; polluted water (rat-tailed maggots)
FOOD:
Adults: Nectar; pollen; aphid honeydew
Larvae: **Predaceous on other insects, particularly aphids; decaying materials;** a few plant feeders

Syrphid fly

Syrphid fly

Syrphid fly larvae feeding on aphids

Flower flies are sometimes referred to as "hover flies" because they are often found hovering near flowers. The adults of many species mimic bees or small yellowjackets, although they do not sting or bite. This common form of mimicry provides protection from predators wary of prey that might sting.

The larvae of some species are very important predators of aphids. The lime green, maggotlike larvae live in or near aphid colonies on leaves, stems, and other plant parts. The toothlike process at the larva's tapered end is used to snare prey and suck the juices from it. Other species produce larvae known as "rat-tailed maggots" because of their long breathing tube. These insects live in polluted water or sometimes in animal carcasses.

Caddisflies

Order: Trichoptera
(try-KAHP-tur-uh)
tricho, hair; *plura,* tail

Characteristics:

SIZE: ~$\frac{1}{2}$ in. (12.8 mm)
SHAPE: Moth shaped
COLOR: **Dull tan**
ANTENNAE: Threadlike; long
MOUTHPARTS: Nonfunctional
EYES: Compound
WINGS: Two pairs; **hairy; held rooflike**
LEGS: Walking
MISCELLANEOUS: **Resemble small, slender moths;** adults nocturnal
LARVAE: **Caterpillar-like; usually live in a silken case**
HABITAT:
Adults: Close to water
Larvae: Lakes or streams
FOOD:
Adults: Liquids or none
Larvae: Decomposing organic matter; predaceous on small aquatic organisms; plant feeders
METAMORPHOSIS: Complete (egg, larva, pupa, adult)

Adult caddisflies are fragile-looking insects with long, thin antennae, and their wings are covered in fine hair and are held rooflike. They resemble tiny moths but do not have the typical coiling proboscis. Their mouthparts are reduced and nonfunctional, and they probably do not eat. Caddisflies are active at night and often attracted to lights.

The adult female caddisfly lays her eggs in water, where the newly emerged larvae build a case from leaves, twigs, and small pebbles held together by silk. The developing larvae crawl along the bottom of shallow, well-oxygenated water, feeding primarily on algae and other microorganisms. Pupation occurs in the same larval case.

Caddisfly larvae are an important food for many fish.

Caddisfly

Adults: Butterflies, Moths, Skippers

Larvae: Caterpillars

Order: Lepidoptera
(lehp-ih-DAHP-tur-uh)
lepido, scale; *ptera,* wing

Characteristics:

WINGSPAN: $\frac{1}{4}$ to >6 in. (6.4 to >150.0 mm)
ANTENNAE: **Threadlike with knob or curved knob, or feathery**
MOUTHPARTS: Adults, siphoning; larvae, chewing
EYES: Compound
WINGS: Two pairs; **typically covered with scales**
LEGS: Walking
LARVAE: **Well-developed head; three pairs of thoracic legs and two to five pairs of fleshy prolegs**
HABITAT: **Plants**
FOOD:
Adults: Nectar
Larvae: Most feed on plants or plant products; very few feed on other insects
METAMORPHOSIS: Complete (egg, larva, pupa, adult)

There are over eleven thousand species of butterflies, moths, and skippers in the United States and Canada. Most are recognizable by their large wings covered with scales that can be easily rubbed off when handled. Some of the most beautiful insects occur in this group, as well as some of the most destructive insect pests.

Butterfly and moth larvae, commonly called "caterpillars," typically have a distinct head, a body with thirteen segments, and a pair of jointed legs on each of the first three segments behind the head. Lepidoptera larvae are unique in having two to five pairs of stubby prolegs with small hooklike structures called crochets, which they use to hang on to the host plants.

Adult lepidopterans use a long, coiled proboscis to siphon nectar, water, or other plant exudates. The larvae, however, have chewing mouthparts and feed almost exclusively on plant parts. Some species feed only on a few plant species, while others have a wide host range. A very few species of Lepidoptera larvae are predaceous on small arthropods.

The order Lepidoptera is often separated into moths, butterflies, and skippers. Although this classification system is not accepted by most lepidopteran taxonomists, it is useful for the more common species found in Texas.

Moths are typically active at night. The females usually have threadlike antennae; the males' antennae are feathery. Their generally drab-colored wings lie flat over their body when at rest. Moths most often pupate in the soil or under tree bark. Butterflies are generally active during the day, are often quite colorful, and hold their wings vertically when at rest. Butterfly antennae are threadlike with a terminal knob. Butterfly pupae, or chrysalids, are generally found aboveground attached to some object by a short stalk called a cremaster.

Skippers combine traits of both moths and butterflies. They are generally active during the daytime, and most hold their wings vertically as butterflies do. However, their bodies are stout and hairy like those of moths. Their antennae are knobbed, like butterfly antennae, but they also possess a slight hook at the end. Skippers pupate in a cocoon formed from leaves or silk.

Common moth families of the order Lepidoptera

Family	*Notable species*	*Larval food source*	*Active time for adults*	*Page*
Bagworms: Psychidae	Evergreen bagworm	Leaves of arborvitae, cedar, other trees	Night	161
Clearwing moths: Sesiidae	Peach tree borer, lesser peach tree borer, squash vine borer	Stems, twigs of trees and vines	Day	162
Ermine moths: Yponomeutidae	Ailanthus webworm	Leaves of the tree-of-heaven	Mornings or evenings	163
Flannel moths: Megalopygidae	Asp, puss caterpillar	Leaves of various trees and shrubs	Night	164
Geometers: Geometridae	Measuringworms, geometers, cankerworms	Leaves of trees and shrubs	Night	165
Lappet moths: Lasiocampidae	Tent caterpillars	Leaves of trees and shrubs	Day	166
Noctuid moths: Noctuidae	Candleflies, millers, cutworms, armyworms	Leaves of numerous plants	Night	167
Royal moths: Saturniidae	Io moth, cecropia moth, luna moth, regal moth, promethea moth, polyphemus moth	Leaves of deciduous trees	Night	168
Sphinx moths, hawk moths, hummingbird moths: Sphingidae	Tomato hornworm, catalpa worms	Leaves of trees, shrubs, vines, and herbs	Day, evening, night	169
Tiger moths: Arctiidae	Woolybears, salt marsh caterpillars	Leaves of numerous plants	Mostly night; a few, day	170

Common butterfly and skipper families of the order Lepidoptera

Families	*Notable groups*	*Larval food*	*Page*
Brush-footed butterflies: Nymphalidae	Anglewings, checker spots, crescents	Leaves of nettle plants and trees	171
Gossamer-wings, hairstreaks: Lycaenidae	Hairstreaks, harvesters, coppers, blues	Internal feeders of flower buds	172
Longwings: Heliconiidae	Fritillaries, heliconians	Passion vines	173
Milkweed butterflies: Danaidae	Monarch butterflies	Milkweeds	174
Satyrs: Satyridae	Browns, satyrs, wood-nymphs	Grasses and sedges	175
Snout butterflies: Libytheidae	American snout	Hackberry	176
Swallowtails: Papilionidae		Trees, shrubs, members of the carrot family	177
Sulphurs, whites, orange-tips: Pieridae		Legumes and mustards	178
Skippers: Hesperiidae		Leaves of numerous plant species	179

Bagworms

Family: Psychidae
(SY-kih-dee)

Characteristics:

WINGSPAN: $\frac{3}{8}$ to 1 in. (9.6 to 25.4 mm)
COLOR: Males, dull black
ANTENNAE: **Feathery; only males have antennae**
MOUTHPARTS: Siphoning
EYES: Compound
WINGS: **Only males have wings; thinly scaled or no scales**
LEGS: Walking
MISCELLANEOUS: **Females wingless, resemble larvae; remain in bag after pupation**
LARVAE: **Live in bag formed from plant materials**
FOOD:
Adults: Nectar
Larvae: **Leaves of red cedar, arborvitae, and other trees**

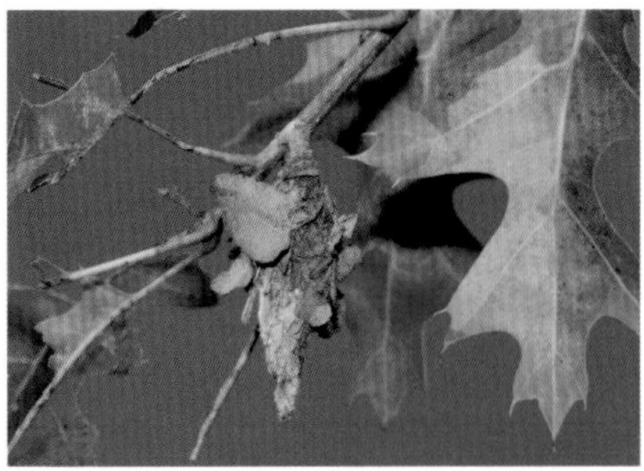

Bagworm (by Gary Brooks)

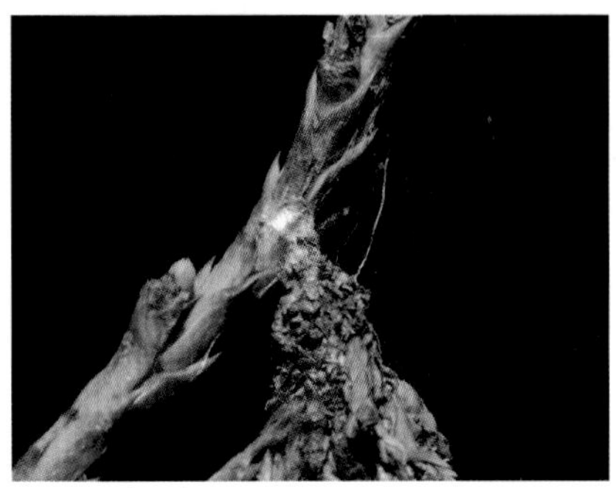

Bagworm band

Bagworm damage

Bagworms are common pests of arborvitae and other trees. The female bagworm, wingless and legless, spends her entire life in a bag made from silk and leaves from the host plant. Winged male bagworms emerge, seek receptive females to mate with, and then die. Male bagworms are seldom seen.

The female remains in the bag until she lays her eggs and dies. The eggs overwinter in the old female bag. In the spring, the larvae hatch and make new bags, which increase in size as the larvae grow. When fully grown, the larvae attach the bag to a branch and pupate.

Damage can result both from larvae-feeding activity and vascular restriction as a result of a band produced to attach the bag to a branch. The evergreen bagworm, *Thyridopteryx ephemeraeformis,* is a common problem on trees in Texas.

Clearwing Moths

Family: Sesiidae
(SEH-see-ih-dee)

Characteristics:

WINGSPAN: $\frac{5}{8}$ to 2 in. (15.9 to 50.8 mm)
COLOR: **Often resembles that of wasps**
ANTENNAE: Threadlike with terminal knob
MOUTHPARTS: Siphoning
EYES: Compound
WINGS: **Mostly without scales and translucent; long and narrow**
LEGS: Walking
MISCELLANEOUS: **Often resemble wasps**

Clearwing moth

FOOD:
Adults: Nectar
Larvae: Plants; bore into roots, stems, vines, and tree branches

Clearwing moths get their name from their clear and almost scaleless wings. Many species are brightly colored and are often mistaken for wasps. All species are active during the day. The sexes are typically different in size and coloration.

The larval stage of the peach tree borer, *Synanthedon exitiosa,* feeds on the sapwood near the soil line of young, nonbearing peach, apricot, and other stone-fruit trees. The feeding activity disrupts the flow of nutrients between the tree's roots and shoots. The larvae of the lesser peach tree borer, *S. pictipes,* are similar to peach tree borer larvae but prefer older trees and are usually found under the bark near wounds.

The larvae of the squash vine borer, *Melittia cucurbitae,* feed inside the stem of squash, gourds, and other members of the cucurbit plant family. The damage is usually not noticed until the vines die beyond the point where the larvae reside.

Ermine Moths

Family: Yponomeutidae
(ee-pahn-uh-MYOO-tih-dee)

Characteristics:

WINGSPAN: $\frac{3}{8}$ to $1\frac{1}{4}$ in. (9.6 to 31.8 mm)
COLOR: **Forewings brightly colored; often with white-and-black patterns**
ANTENNAE: Feathery
MOUTHPARTS: Siphoning
EYES: Compound
WINGS: **Held flat over body; narrow; brightly colored**
LEGS: Walking
MISCELLANEOUS: **Slender moths**

FOOD:
Adults: Nectar
Larvae: Leaves of the tree-of-heaven and other plants

The ailanthus webworm, *Atteva punctella,* is the most common member of the Yponomeutidae in Texas. The adults are active during the daytime, especially mornings and evenings, and are readily recognized by their orange wings with white dots encircled by black lines. The larvae of the ailanthus webworm feed in groups under silken webs on the leaves of ailanthus (also known as the tree-of-heaven) and other plants.

Other species may be differently colored, but most have brightly patterned forewings.

Ermine moth

Adults: Flannel Moths

Larvae: Puss Caterpillars, Asps

Family: Megalopygidae
(mehg-uh-loh-PIHJ-ih-dee)

Characteristics:

WINGSPAN: ½ to 1½ in. (12.8 to 38.1 mm)
COLOR: Brownish or yellowish
ANTENNAE: Feathery
MOUTHPARTS: Siphoning
EYES: Compound
WINGS: **Forewing scales mixed with crinkled, hairlike scales**
LEGS: Walking
LARVAE: **Stout and hairy; some hairs forming a crest down top; stinging hairs mixed with other hairs; seven pairs of prolegs**
FOOD:
Adults: Nectar
Larvae: Typically tree leaves: hackberry, oak, maple, sycamore

Puss caterpillar

The puss caterpillar or asp, *Megalopyge opercularis,* is a common stinging caterpillar found on a variety of trees throughout Texas but most commonly in Central and South Texas. The yellowish and very hairy adults are rarely seen.

These tan, sluglike caterpillars possess many short, stout hairs. Beneath these hairs are hollow spines attached to poison glands. When touched, these urticating spines break open; the toxin oozes out onto the victim and causes dermatitis.

Puss caterpillars overwinter as pupae on twigs in a tough silken cocoon that has a distinct lid.

Flannel moth (by Gary Brooks)

Adults: Geometers

Larvae: Measuringworms, Inchworms, Loopers, Cankerworms

Family: Geometridae
(jee-oh-MEHT-rih-dee)

Geometrid moth

Characteristics:

WINGSPAN: ~1 in. (2.5 cm)
COLOR: Variable; **most white, yellow, brown, or lime green**
ANTENNAE: Variable; sometimes feathery
MOUTHPARTS: Siphoning
EYES: Compound
WINGS: **Broad; forewings marked with wavy lines that often continue on the hind wings**
LEGS: Walking; three distinct pairs
MISCELLANEOUS: Slender bodied
LARVAE: **Two pairs of prolegs at posterior end but none in central portion of body; when disturbed, often stand erect, resembling a twig; often smooth and hairless**
FOOD:
Adults: Nectar
Larvae: Plant leaves

Adult geometrids are typically white, yellow, brown, or lime green moths with diagonal wavy lines on their wings. Geometrids are most active at night; many are attracted to lights.

The larvae of geometrids typically have no prolegs in the center section of the abdomen. As the larvae crawl, the caterpillar grasps the surface with the front thoracic legs and arches its back, bringing its rear prolegs forward. The rear prolegs cling to the surface as the body is stretched forward to repeat the cycle. The names "inchworm," "loopers," and "measuringworms" reflect this unique form of locomotion.

The larvae of many species resemble twigs. When disturbed, the caterpillar stands erect and motionless, mimicking its surroundings.

Measuringworm

Adults: Lappet Moths

Larvae: Tent Caterpillars

Family: Lasiocampidae
(lay-see-oh-KAM-pih-dee)

Characteristics:

WINGSPAN: 1 to 3 in. (2.5 to 7.6 cm)
COLOR: Brown or gray; wings typically banded
ANTENNAE: Somewhat feathery in both sexes
MOUTHPARTS: Absent
EYES: Compound
WINGS: With scales
LEGS: Walking
MISCELLANEOUS: Body, legs, and eyes are hairy
LARVAE: **Generally brown; hairy; often gregarious**
FOOD:
Adults: Do not eat
Larvae: **Tree leaves**

Tent caterpillars

The eastern tent caterpillar, *Malacosoma americanum,* and the forest tent caterpillar, *M. disstria,* are common pests of shade trees in Texas. Eastern tent caterpillars are gregarious caterpillars that produce a silken web away from the leaf food source and in the fork of tree limbs. During the day, the caterpillars leave the protective web and seek food.

The forest tent caterpillars are also gregarious but do not form a silken web. These caterpillars are easily recognized by the keyhole-shaped markings on their backs.

Adult tent caterpillars are generally brown or gray and active at night. Females lay eggs in late summer in a mass that encircles small twigs. There is typically one generation per year, and they overwinter in the egg stage.

Forest tent caterpillar, *Malacosoma disstria*

Adults: Noctuid Moths

Larvae: Armyworms, Cutworms, and others

Family: Noctuidae
(nahk-TOO-ih-dee)

Black witch, *Ascalapha odorata* (by Jack Brady)

Characteristics:

WINGSPAN: $\frac{7}{8}$ to 6 in. (2.3 to 15.2 cm)
COLOR: **Typically dull gray or brown; sometimes mottled**
ANTENNAE: Slender and threadlike
MOUTHPARTS: Siphoning
EYES: **Compound; glow when struck by light**
WINGS: **Folded rooflike at rest; often with spots or lines on wings**
LEGS: Walking
MISCELLANEOUS: **Attracted to lights and often seen on porches and other well-lighted areas**
LARVAE: Smooth and dull colored; typically five pairs of prolegs (loopers only three pairs)
FOOD:
Adults: Nectar
Larvae: Plant roots, shoots, and fruit

The noctuids are the largest family of Lepidoptera with over twenty-seven hundred species in North America. These moths are typically dull colored, active at night, and often attracted to light. They are sometimes referred to as "candleflies" because of their attraction to firelight.

Noctuids represent one of the most destructive groups of insects. Numerous species of armyworms, cutworms, and loopers attack a wide range of crops and ornamental plants. Some species have a very narrow host range, while others are quite broad.

The black witch, *Ascalapha odorata,* is one of the largest moths in Texas, with a wingspan approaching six inches.

Noctuid moth

Armyworm

Adults: Royal Moths

Larvae: Giant Silkworms

Family: Saturniidae
(sa-tur-NEE-ih-dee)

Characteristics:

WINGSPAN: 1 to 6 in. (2.5 to 15.2 cm)
COLOR: **Many brightly colored; often with bold markings**
ANTENNAE: Feathery about halfway up the antennae; **males much broader than females**
MOUTHPARTS: Reduced; adults do not feed
EYES: Compound
WINGS: **Often with transparent eyespots; some with tails on hind wings**
LEGS: Walking
MISCELLANEOUS: **Densely hairy body**
LARVAE: **Large; often with spines, some stinging**
FOOD:
 Adults: Do not feed
 Larvae: **Leaves of deciduous trees and shrubs**

The saturniids are the largest and most spectacular moths in North America. Their size, wing markings, and the presence of eyespots on the wings are generally enough to make an accurate determination of this family. There are about sixty-eight species in the United States and Canada. The regal moth, *Citheronia regalis;* polyphemus moth, *Antheraea polyphemus;* luna moth, *Actias luna;* and io moth, *Automeris io* are quite common in Texas. Most prefer wooded areas.

Polyphemus moth, *Antheraea polyphemus*

The larvae usually feed on deciduous trees shrubs and occasionally can remove large portions of the leaves, but they seldom cause economic damage. Adult saturniids' mouthparts are reduced, and they do not feed. Males and females can be similar or quite different depending on the species, but the males can generally be distinguished by their larger, broader antennae. These moths generally are active at night and are readily attracted to lights.

The larvae are usually greenish. Members of one genus, the io moths, have stinging spines. Saturniids usually pupate in a silken cocoon, and silk from some Asian species is commercially produced.

Luna moth, *Actias luna*

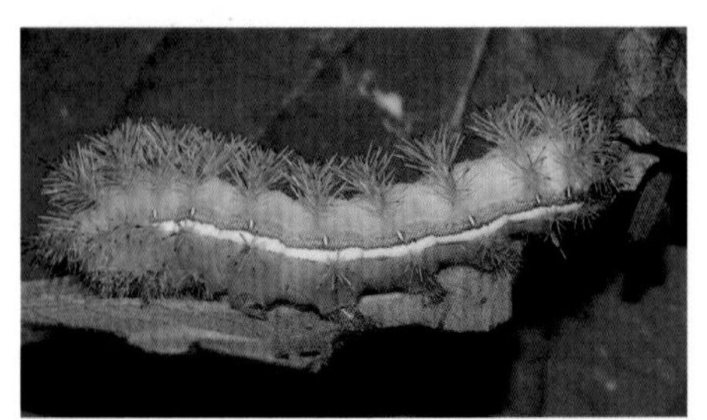

Io larva, *Automeris io*

Adults: Sphinx Moths, Hawk Moths, Hummingbird Moths

Larva: Hornworms

Family: Sphingidae
(SFIHN-jih-dee)

Sphinx moth

Characteristics:

WINGSPAN: **$2\frac{1}{2}$ to 5 in. (5.1 to 12.7 cm)**
COLOR: Variable
ANTENNAE: **Feathery;** often thickened in the middle
MOUTHPARTS: **Long proboscis**
EYES: Compound
WINGS: **Narrow; pointed; hind wings smaller than forewings**
LEGS: Walking
MISCELLANEOUS: **Cigar-shaped body; fast fliers; some species can hover**
LARVAE: Typically large and green; **often possess a horn at apex of abdomen;** sometimes with distinct pattern on back
FOOD:
Adults: Nectar
Larvae: Leaves of shrubs, vines, and trees

Sphinx moths are common moths frequently seen at dusk, hovering near flowers as they siphon nectar. This is a rare feat in the insect world; they are often referred to as "hummingbird moths."

Adult sphinx moths are easy to identify by their cigar-shaped body, narrow and pointed wings, and long proboscis. Some species resemble bumble bees, having little or no scale on their wings, and are known as "clearwing sphinx moths." Their appearance is probably defensive mimicry.

The larvae can be over three inches (7.6 cm) long when fully grown, are typically green, and have a spinelike projection at the base of the abdomen. Although this spine is harmless, potential predators (and humans) may not know this and retreat rather than risk being stung. Sphinx moths usually pupate in a cell in the soil with no cocoon.

The tomato hornworm, *Manduca quinquemaculata,* is a common pest of tomatoes, potatoes, and tobacco.

Clearwing sphinx moth

Hornworm

Adults: Tiger Moths

Larvae: Woolybears

Family: Arctiidae
(ark-TEE-ih-dee)

Characteristics:

WINGSPAN: $\frac{3}{4}$ to 3 in. (1.9 to 7.6 cm)
COLOR: **White, yellow, orange, or red with black spots or stripes on forewings**
ANTENNAE: Feathery
MOUTHPARTS: Siphoning
EYES: Compound
WINGS: Held rooflike when at rest; **often spotted or striped**
LEGS: Walking
LARVAE: **Usually very hairy; some times called woolybears**
FOOD:
Adults: Nectar
Larvae: Typically leaves of grasses and herbaceous weeds; occasionally trees and shrubs

Tiger moths are usually easy to recognize by their striped or spotted wings, which give rise to their common name. Most adults are active at night and attracted to lights. However, members of the subfamily Ctenuchinae are active during the day and may resemble wasps. The yellowcollared scape moth, *Cisseps fulvicollis,* is dark colored with an orange collar.

Most larvae have long hairs and are often referred to as "woolybears" or "woolyworms." Most species are nocturnal, feed on grasses or weeds, and are generally of no economic importance. The fall webworm, *Hyphantria cunea,* however, is a notable exception. These larvae live in groups within silken webs that enclose limbs of trees. The unsightly tent and damaged leaves are a common sight in shade trees during late summer.

Salt marsh caterpillar

Tiger moth

Lichen moth

Ctenucha moth

Brush-footed Butterflies

Family: Nymphalidae
(nihm-FAL-ih-dee)

Characteristics:

WINGSPAN: $1\frac{1}{5}$ to 3 in. (3.0 to 7.6 cm)
COLOR: Orange, brown, or black
ANTENNAE: Threadlike with terminal knob
MOUTHPARTS: Siphoning
EYES: Compound; hairy
WINGS: **Many have crescent-shaped spots on wings**
LEGS: **Forelegs reduced; appear to have only two pairs**
LARVAE: **Spiny; some resemble bird droppings**
FOOD:
Adults: Nectar; some males feed on dung, carrion, and mud
Larvae: Variable

This large, very diverse, and common group of butterflies includes the anglewings, fritillaries, crescents, and others. The family's distinguishing characteristic is the reduced forelegs that suggest these butterflies have only two pairs of legs, thus their common name. Many authorities place these butterflies as a subfamily Nymphalinae within the family Nymphalidae.

Anglewing

The viceroy, *Limenitis archippus,* is very similar to the monarch butterfly (order Lepidoptera, family Danaidae) but is differentiated by the black line across the hind wing. Because the monarch accumulates toxic compounds from the milkweed plant it feeds on, most predators quickly learn to avoid it. The viceroy does not possess these toxins but benefits from its close resemblance to the monarch.

The larvae are also quite variable, but most species possess spines that offer protection from predators. Other larvae are cryptically colored and blend in with their surroundings.

Mourning cloak, *Nymphalis antiopa*

Buckeye, *Junonia coenia*

Goatweed leafwing, *Anaea andria* (by Philip Adams)

Gossamer-wings, Hairstreaks

Family: Lycaenidae
(ly-SIHN-ih-dee)

Characteristics:

WINGSPAN: ¾ to 2 in. (1.9 to 5.1 cm)
COLOR: **Often brightly colored; metallic markings on wings**
ANTENNAE: Threadlike and knobbed; **usually ringed with white**
MOUTHPARTS: Siphoning
EYES: Compound; **white scales encircling eyes**
WINGS: **Often with threadlike tails on hind wings**
LEGS: **Forelegs reduced in males**
MISCELLANEOUS: **Rest with wings closed and held over body**
LARVAE: **Sluglike; variable colors**
FOOD:
Adults: Nectar
Larvae: **Plant flower buds; rarely predaceous on aphids**

Gossamer-wings are a large group of small, colorful butterflies. The family is usually divided into four subfamilies: harvesters, blues, coppers, and hairstreaks. Some authorities also include the metalmark butterflies.

Only one species of harvester occurs in the United States; it can be occasionally found near Houston and southward. It is unique in that the larvae are predaceous, feeding on wooly aphids. Coppers, as the name implies, are a coppery orange color and are rarely found in Texas. Blues are small, fragile butterflies whose upper surface of the forewings is blue, particularly in the males.

Hairstreaks are the most common gossamer-wing butterflies in Texas. They are typically gray or brown with two or three hairlike tails on the hind wings. When resting, hairstreaks constantly move their hind wings, creating the impression that the tails are actually antennae. Hairstreaks hope that would-be predators will attack the wrong end and give it enough time to escape.

The larvae of gossamer-wings are distinctly sluglike and quite hairy. Most species feed inside flower buds, leaving their posterior outside the flower structure. Many species have a mutually beneficial relationship with ants. The ants milk the caterpillar for a liquid produced by a specialized gland. In return, the ants protect the larvae from potential predators.

Hairstreak

Hairstreak (by Gary Brooks)

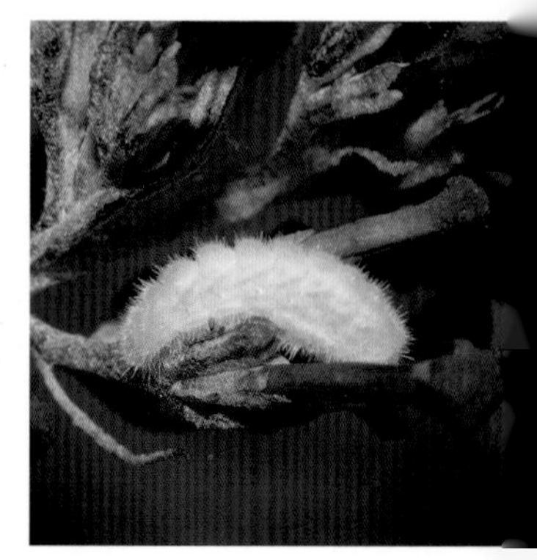
Hairstreak larva (by Gary Brooks)

Longwings

Family: Heliconiidae
(heh-lih-kahn-EE-ih-dee)

Characteristics:

WINGSPAN: $2\frac{1}{2}$ to 4 in. (6.4 to 10.2 cm)
COLOR: **Reddish orange or black with yellow stripes**
ANTENNAE: Threadlike with club
MOUTHPARTS: Siphoning
EYES: Compound
WINGS: **Relatively long and narrow; some with silver markings on underside of hind wing**
LEGS: **Forelegs reduced; appear to have only two pairs**
LARVAE: Dark with branching spines
FOOD:
- Adults: Nectar from composite flowers; lantana
- Larvae: **Various species of passion vines**

Butterflies of this group have reduced forelegs and appear to have only four legs. This group is often assigned to the subfamily Heliconiinae under the family Nymphalidae. Heliconiids can generally be separated from other nymphalids by their long, narrow wings.

Gulf fritillary, *Agraulis vanillae* (by Gary Brooks)

This small group consists of about five species in the United States, but the gulf fritillary, *Agraulis vanillae,* is a common butterfly throughout Texas. It has orange wings with black-rimmed white spots on the upper side. The undersides of its wings reveal beautiful silvery streaks.

The zebra butterfly, *Heliconius charitonius,* has long, narrow wings that are black with narrow, yellow stripes giving a distinctly zebralike appearance. These butterflies are rarely found north of the Houston area.

The larvae of this group feed on passion vines and probably accumulate chemical compounds that render the larvae and adults distasteful to most predators.

Gulf fritillary, *Agraulis vanillae* (by Philip Adams)

Milkweed Butterflies

Family: Danaidae
(duh-NAY-ih-dee)

Characteristics:

WINGSPAN: 3½ to 4 in. (8.9 to 10.2 cm)
COLOR: **Typically brown with black-and-white markings**
ANTENNAE: Threadlike with terminal knob
MOUTHPARTS: Siphoning
EYES: Compound
WINGS: **Light brown; bordered in black with white dots**
LEGS: **Forelegs reduced; appear to have only two pairs**
LARVAE: **Smooth; brightly colored; long fleshy filaments on back**
FOOD:
Adults: **Nectar from milkweed and other plants**
Larvae: **Milkweeds**

Monarch butterfly, *Danaus plexippus*

Queen, *Danaus gilippus* (by Joe Carter)

Monarch larva

Milkweed butterflies have reduced forelegs, so they appear to have only four legs. This group is often assigned to the subfamily Danainae within the family Nymphalidae.

The monarch butterfly, *Danaus plexippus,* is one of the most common and well-known butterflies in Texas. Its brown wings have a black border with two rows of white dots.

The larvae of the monarch butterfly feed on milkweed plants, which are toxic to most animals. These chemicals accumulate in the immature insect and remain through the developmental process to the adult stage. Predators, such as birds, soon associate monarchs with an unpleasant taste and learn to avoid them.

Several other butterfly species, which lack distasteful chemicals, mimic monarchs and achieve some level of protection as a result of this trickery.

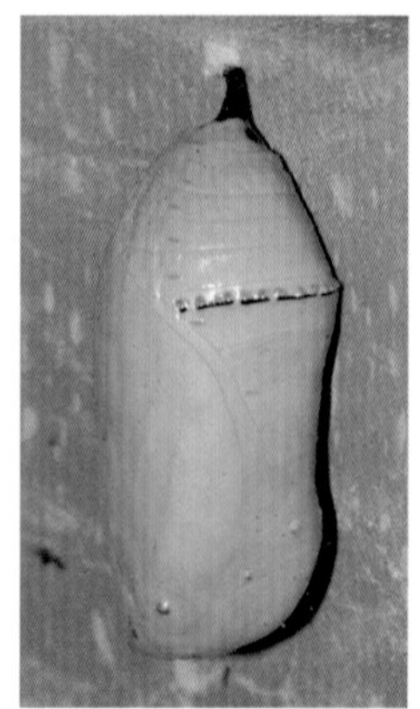

Monarch pupa

Satyrs

Family: Satyridae
(suh-TEER-ih-dee)

Characteristics:

WINGSPAN: $1\frac{1}{2}$ to 4 in. (3.8 to 10.2 cm)
COLOR: **Brownish or grayish**
ANTENNAE: Threadlike with terminal knob
MOUTHPARTS: Siphoning; short proboscis
EYES: Compound
WINGS: **Generally with eyespots on wing margins**
LEGS: **Forelegs reduced; appear to have only two pairs**
MISCELLANEOUS: **Adults fly in a characteristic bobbing back-and-forth motion**
LARVAE: **Forked tails**
FOOD:
Adults: **Rotting fruit; animal droppings; plant sap**
Larvae: **Grasses; sedges**

Butterflies of this group have reduced forelegs, so they appear to have only four legs. This group is often assigned to the subfamily Satyrinae under the family Nymphalidae. Adult satyrids have short proboscises and feed on rotting fruit or manure rather than flower nectar.

The larvae of satyrids are typically smooth with short hairs and forked tails and often possess hornlike projections on their heads. Most larvae feed on grasses or sedges.

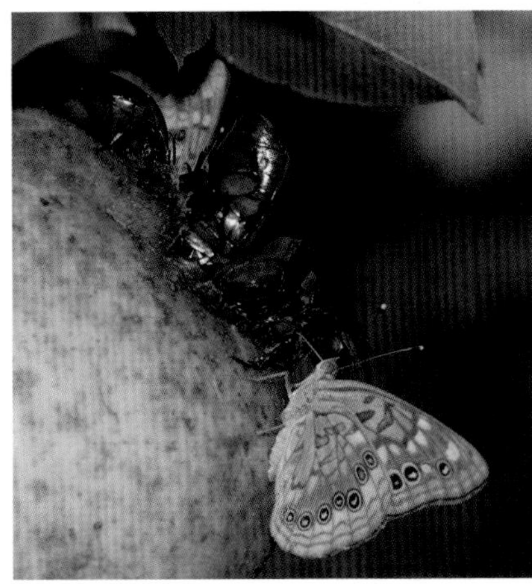

Satyr feeding on rotting pear

Satyr (by Gary Brooks)

Snout Butterflies

Family: Libytheidae
(lih-bih-THEE-ih-dee)

American snout, *Libytheana carinenta*

Characteristics:

WINGSPAN: ~$1\frac{3}{4}$ in. (4.4 cm)

COLOR: **Brownish with orange or white markings**

ANTENNAE: Threadlike with club

MOUTHPARTS: Siphoning; **palps long and projecting forward**

EYES: Compound

LEGS: Females, six visible legs; males, forelegs reduced; appear to have only two pairs

WINGS: **Notched; undersides resemble dead leaves**

LARVAE: **Smooth; yellow green with yellow stripe**

FOOD:

Adults: Nectar from a variety of plants

Larvae: **Hackberry**

Snout butterflies have reduced forelegs, so they appear to have only four legs. This group is often assigned to the subfamily Libytheinae under the family Nymphalidae.

This is a small group with only the American snout, *Libytheana carinenta,* commonly found in the United States. This butterfly is easily recognized by the long forward-projecting palps that form a snout, inspiring its common name. Its wings are deeply notched, and the undersides resemble a dead leaf, which offers protection from predation.

The larvae feed on hackberry leaves.

American snout, *Libytheana carinenta*

Swallowtails

Family: Papilionidae
(pa-pihl-ee-AHN-ih-dee)

Characteristics:
WINGSPAN: 2 to 5 in. (5.1 to 12.7 cm)
COLOR: **Variable; often colorful**
ANTENNAE: Threadlike with knob
MOUTHPARTS: Siphoning
EYES: Compound
WINGS: **Usually one or more tail-like projections on the hind wings**
LEGS: Three fully developed pairs
LARVAE: Smooth bodied; **some cryptically colored; possess osmeterium**
FOOD:
Adults: Nectar
Larvae: Leaves of trees, shrubs, carrots, parsley, and other plants

Giant swallowtail, *Papilio cresphontes* (by Philip Adams)

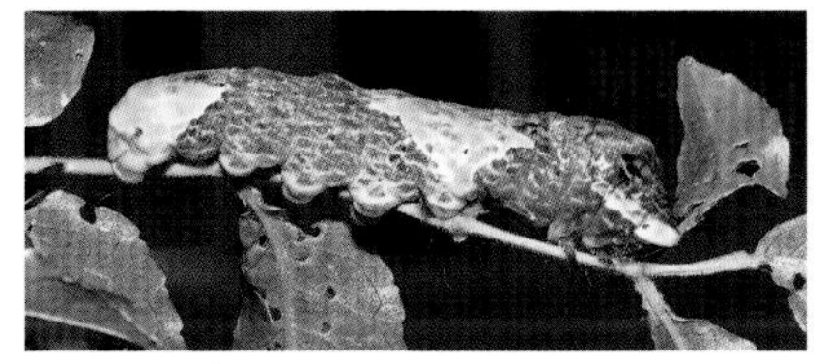

Swallowtail larva (by Gary Brooks)

Zebra swallowtail, *Eurytides narcellus* (by James Lasswell)

Swallowtails are among our largest and most beautiful butterflies. The adults are generally colorful and usually possess distinctive, tail-like projections on their hind wings.

The larvae are smooth and have a fleshy, retractable, hornlike structure located behind the head, called an osmeterium. When disturbed, the larva displays this organ, which emits a rather pungent odor.

Early stages of swallowtail larval development have another interesting defensive mechanism. Most species bear a striking resemblance to bird droppings. These are the ugly ducklings of the insect world because the rather unattractive larvae transform into some of our most beautiful butterflies.

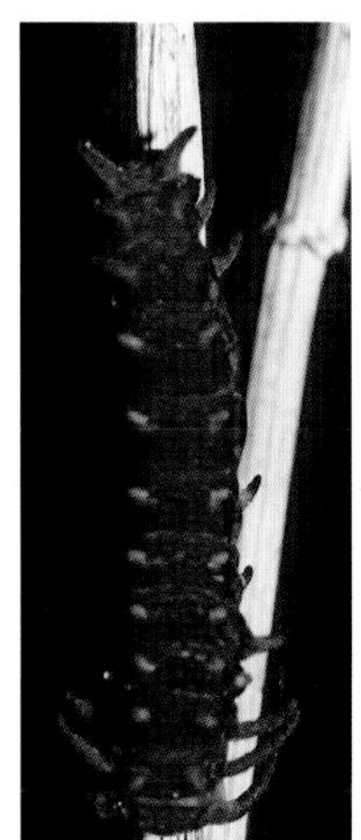

Pipevine swallowtail larva, *Battus philenor*

Common Texas species include the black swallowtail, *Papilio polyxenes;* the tiger swallowtail, *P. glaucus;* and the pipevine swallowtail, *Battus philenor.*

Sulphurs, Whites, Orangetips

Family: Pieridae
(py-EHR-ih-dee)

Characteristics:

WINGSPAN: $1\frac{1}{2}$ to $2\frac{1}{2}$ in. (3.8 to 6.4 cm)
COLOR: **Generally white, orange, or yellowish; black marginal wing markings**
ANTENNAE: Threadlike and knobbed
MOUTHPARTS: Siphoning
EYES: Compound
WINGS: **Compact; rounded and generally solid color with sparse dark markings**
LEGS: Three pairs of fully developed legs
LARVAE: Smooth; cylindrical
FOOD:
Adults: Nectar
Larvae: Members of the legume and mustard plant families

Sulphur (by Joe Carter)

Pierids are a common group of butterflies in Texas. These medium-sized butterflies are predominantly solid colors but typically have dark markings on the wing margins. The group is often separated into three subfamilies: whites, orangetips, and sulphurs. These butterflies typically fly in a straight line and are often observed in large groups.

The larvae of pierids are typically smooth but with short hairs. They are generally solitary feeders, preferring plants of the mustard or legume families. A few species can cause economic damage to crops.

Sulphurs (by Joe Carter)

White

Skippers

Family: Hesperiidae
(hehs-pehr-EE-ih-dee)

Characteristics:

WINGSPAN: 1 to 2 in. (2.5 to 5.1 cm)
COLOR: **Typically drab**
ANTENNAE: **Widely separated at base; generally hooked at end**
MOUTHPARTS: Siphoning
EYES: Compound
WINGS: **Forewings and hind wings held at different angles when at rest; some species with tails**
LEGS: Three pairs of fully developed legs
MISCELLANEOUS: Fast and erratic fliers; stout bodied
LARVAE: **Smooth; large head and restricted neck**
FOOD:
Adults: Nectar
Larvae: Plant leaves; usually feed inside a leaf shelter; oaks, legumes, grasses

Skipper

Skippers get their name from their fast and erratic flight pattern and have characteristics of both moths and butterflies. Skippers normally have stout, hairy bodies much like those of moths. However, skippers hold their wings upright and are generally active during the day. Their antennae are more like those of the butterflies but have a distinctive hook at the tip.

The larvae of skippers have a large head and narrow neck region. Their bodies are smooth and may be covered with short hairs. Most live in folded leaves and feed only at night but generally do not cause noticeable damage.

Duskywing, *Erynnis* spp. (by Joe Carter)

Long-tailed skipper (by Philip Adams)

Wasps, Ants, Bees

Order: Hymenoptera
(hy-mehn-AHP-tur-uh)
hymen, membrane; *ptera,* wing

Characteristics:

SIZE: Minute to >2 in. (5.1 cm)
ANTENNAE: Typically threadlike; a few elbowed
MOUTHPARTS: Modified for chewing, lapping, or cutting
EYES: Compound
WINGS: Two pairs; **hind wings smaller than forewings; typically translucent; occasionally absent**
LEGS: Typically for walking
MISCELLANEOUS: **Often social and form colonies; many possess a sting**
HABITAT:
Adults and Larvae: Variable
FOOD: A few plant feeders; a few scavengers; many parasitoids; a few predators
METAMORPHOSIS: Complete (egg, larva, pupa, adult)

Hymenoptera is a large and very diverse group of insects. Many are quite common, such as bees, ants, paper wasps, and mud daubers; others are rarely seen. Some of the smallest insects known are in this group. Some species of tiny wasps live and develop inside the eggs of other insects.

Most species of Hymenoptera have two pairs of translucent wings that possess tiny hooks, called hamuli, which connect the forewings and hind wings during flight. Some species, however, are wingless.

Many species of Hymenoptera are beneficial because they feed on and regulate the numbers of other insects. The females of beneficial wasp species typically lay their eggs on or in the host, and the developing larvae feed on the host insects without initially killing them. When the wasp completes its development, the adult emerges, and the host insect is killed.

Other species of Hymenoptera are very important plant pollinators. Many crops, including cotton, peanuts, legumes, and fruit trees, are dependent on or benefit from insect pollination. Hymenoptera also have the dubious honor of being the only insect group that may have a sting. This structure is actually the modified ovipositor of females and used both to paralyze prey and to defend itself against predators. Male hymenopterans cannot sting.

Many species of Hymenoptera have developed complex social systems. Some bees and ants have elaborate caste systems composed mainly of infertile female workers, one to a few reproductive queens, and very few males. In some species asexual reproduction is the norm, and males are only rarely found.

COMMON FAMILIES OF THE ORDER HYMENOPTERA, WITH THEIR PREFERRED HABITATS AND FEEDING CHARACTERISTICS

Family	*Colonies*	*Solitary*	*Ground dwellers*	*Plants*	*Predators or parasitoids*	*Page*
Ants: Formicidae	X	X	X	X	X	182
Bees: Apidae	X	X	X	X		183
Braconid wasps: Braconidae		X		X	X	184
Chalcids and others: superfamily Chalcidoidea		X		X	X	185
Cuckoo wasps: Chrysididae		X		X	X	186
Gall wasps: Cynipidae		X		X		187
Ichneumon wasps: Ichneumonidae		X			X	188
Mud daubers: Sphecidae		X			X	189
Leaf-cutting bees: Megachilidae		X		X		190
Paper wasps, yellowjackets, hornets: Vespidae	X	X (few)	X		X	191
Scoliid wasps: Scoliidae		X			X	192
Spider wasps: Pompilidae		X	X		X	193
Stephanids: Stephanidae		X		X	X	194
Sweat bees: Halictidae		X		X	X	195
Tiphiid wasps: Tiphiidae		X			X	196
Velvet ants, cow killer wasps: Mutillidae		X	X		X	197

Ants

Family: Formicidae
(for-MIHS-ih-dee)

Characteristics:

SIZE: $\frac{1}{16}$ to $\frac{1}{2}$ in. (1.6 to 12.8 mm)
SHAPE: **Abdomen attached to thorax by thin waist with one or two nodes**
COLOR: Generally solid red, black, or brown
ANTENNAE: **Elbowed;** 6- to 13-segmented
MOUTHPARTS: Chewing
EYES: Compound
WINGS: **Both winged and wingless forms**
LEGS: Walking
HABITAT: Variable; soil; plants
FOOD:
Adults: Scavengers; predators of other small arthropods; plant feeders
Larvae: Food prepared by workers

Elbowed antennae

Ants are one of the most commonly observed animals on earth. They are easily recognized by their thin waist and elbowed antennae. Most species form a complex caste system composed of winged queens and kings and sterile, wingless female workers.

The winged queens and kings often appear shortly after a spring rain, mate, establish new colonies, and lose their wings. The males are no longer needed by the colony and soon die, leaving only the queen and sterile female workers.

Harvester ants, *Pogonomyrmex barbatus,* a favorite meal for hungry horned lizards, *Phrynosoma* spp., are common in open fields, where they make large, circular areas void of vegetation around their nest opening. The imported fire ant, *Solenopsis invicta,* was accidentally introduced to Alabama in the 1930s from ships coming from South America. Since their arrival, fire ants have spread across much of the southern United States.

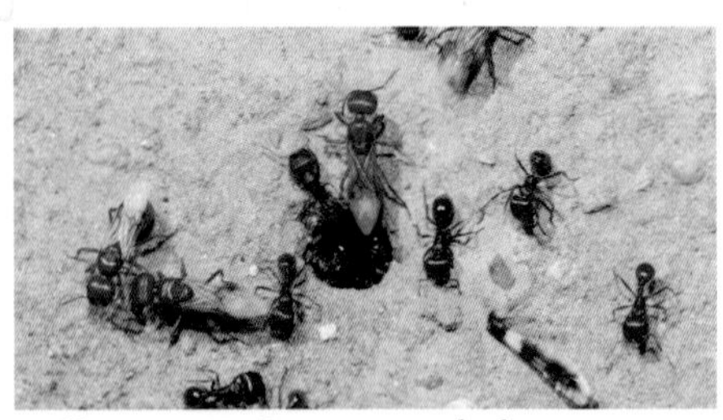

Harvester ants, *Pogonomyrmex barbatus*

Imported fire ant, *Solenopsis invicta,* mound (by Gary Brooks)

Ants feeding on dead grasshopper

Bees: Honey Bees, Bumble Bees, Carpenter Bees, Digger Bees

Family: Apidae
(AY-pih-dee)

Honey bee, *Apis mellifera*

Characteristics:

SIZE: $\frac{5}{8}$ to $1\frac{1}{16}$ in. (1.6 to 2.7 cm)

SHAPE: Typically beelike

COLOR: Body black; generally with yellow markings

ANTENNAE: **Elbowed**

MOUTHPARTS: **Modified into a long, flattened tongue for lapping nectar**

EYES: Compound; **hair between facets**

WINGS: Present

LEGS: Walking; **hind tarsi enlarged and serve as a pollen sac**

MISCELLANEOUS: **Body hairs thick and branched;** adults generally diurnal

HABITAT:
- Adults: Flowers
- Larvae: Hive

FOOD:
- Adults: Nectar
- Larvae: Food prepared by workers

Apidae includes the well-known honey bees, *Apis mellifera,* and bumble bees, *Bombus* spp. These insects form colonies composed of a queen, sterile female workers, and, on occasion, males. Honey bees are considered among the most important pollinators of many crops and fruit trees. They build hives in aboveground cavities, such as tree holes and wall voids. Bumble bees, however, typically nest underground in abandoned rat holes or other soil cavities.

In addition to these common social insects, digger bees, cuckoo bees, and carpenter bees are members of this family. These insects are solitary and nest in the ground or, in the case of carpenter bees, in cells built in wood by the adults.

Bumble bee, *Bombus* spp. (by Joe Carter)

Wild honey bees

Braconid Wasps

Family: Braconidae
(bruh-KAH-nih-dee)

Apanteles spp. pupae on caterpillar (by Gary Brooks)

Characteristics:

SIZE: ⅛ to ½ in. (3.2 to 12.8 mm)
SHAPE: Typically wasplike
COLOR: **Generally brown or black**
ANTENNAE: Threadlike; 16-segmented
MOUTHPARTS: Chewing
EYES: Compound
WINGS: Present
LEGS: Walking
MISCELLANEOUS: **Resemble ichneumons;** ovipositor arises in front of apex of abdomen and is often longer than body
HABITAT:
- Adults: Free living
- Larvae: Solitary and gregarious; primary and secondary parasitoids

Braconid wasp

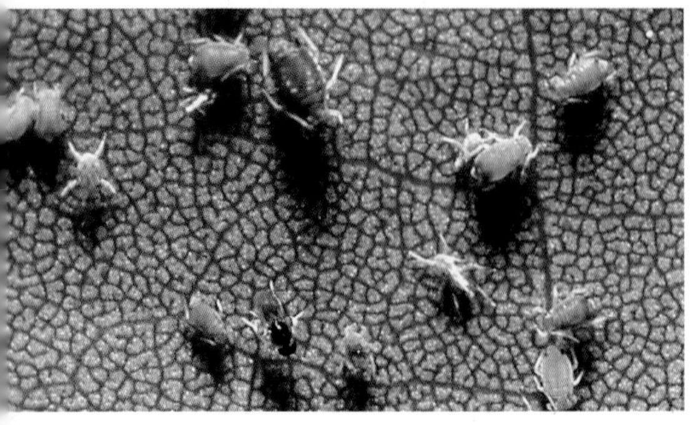

Aphid parasitoid

FOOD:
- Adults: Nectar; ooze from ovipositor wound on host
- Larvae: Ectoparasitoids or endoparasitoid of other insects

Braconids are a large and very diverse group with no easily distinguishable characteristics. They are usually brown or black and have antennae that are typically at least one-half the length of the body. The female's ovipositor is often longer than the body. Braconids are similar to the ichneumons (order Hymenoptera, family Ichneumonidae) and are often difficult to distinguish, particularly in the field.

Most braconids are either ecto- or endoparasitoids of beetles, caterpillars, and some hemipterans and are important in keeping many pest species in check. They may be solitary or gregarious, primary or secondary parasitoids, and host specific or generalist. Often the larval stage emerges from the host and spins a cocoon for pupation. As the host's body collapses, the cocoons remain and resemble a cluster of cotton balls.

Chalcids

Superfamily: Chalcidoidea
(chal-sih-DOY-dee-uh)

Characteristics:

SIZE: **Very small; < $\frac{1}{10}$ in. (2.5 mm)**
SHAPE: **Extremely variable**
COLOR: Generally black, blue black, or greenish; many metallic
ANTENNAE: **Elbowed;** 5- to 13-segmented
MOUTHPARTS: Chewing
EYES: Compound
WINGS: Venation greatly reduced; wings held flat over abdomen when at rest
LEGS: Walking; **hind legs often enlarged and toothed**
HABITAT:
 Adults: Free living
 Larvae: Host
FOOD:
 Adults: Plant sap; honeydew
 Larvae: Parasitoids, hyperparasitoids of other insects; a few phytophagous

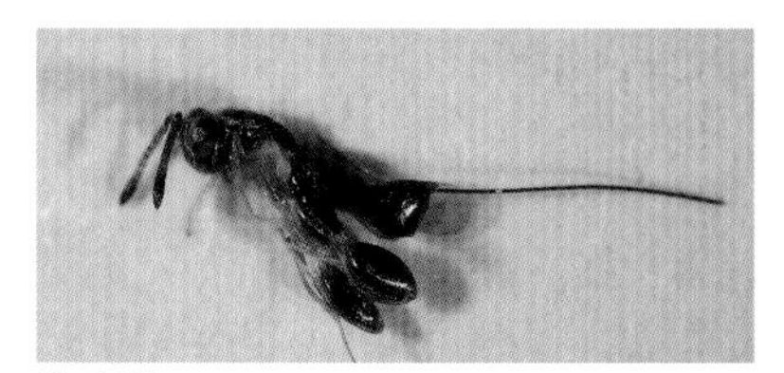

Chalcid wasp

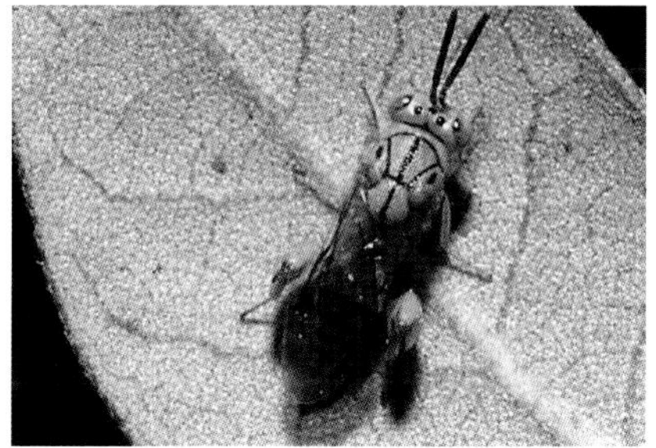

Chalcid wasp

Chalcid wasp

The Chalcidoidea superfamily is a group of about eighteen families of tiny wasps. Most species are less than one-tenth inch and a few, less than four-thousandths inch, long. To visualize their size, consider that the members of several families are parasitoids that complete their development inside and emerge as adults from the eggs of other insects.

Most families within this group are primary endo- or ectoparasitoids. They are beneficial insects, attacking a wide range of hosts, including all stages of development of insects, mites, and even spider eggs. However, a few species are secondary or tertiary parasitoids that attach to the parasitoid feeding on the primary host insect. A few species are predators as larvae, and some are gall formers.

Members of the genus *Encarsia* (family Aphelinidae) are important parasitoids of whiteflies (order Hemiptera, family Aleyrodidae) and scale insects (order Hemiptera, family Coccidae). The egg parasitoid *Trichogramma* spp. can be purchased and released for biocontrol of pests on farms and in the garden.

Chalcid egg parasitoids

Cuckoo Wasps

Family: Chrysididae
(kry-SIH-dih-dee)

Characteristics:

SIZE: $\frac{1}{4}$ to $\frac{1}{2}$ in. (6.4 to 12.8 mm)
SHAPE: **Beelike; abdomen hollowed out ventrally; last segment of abdomen toothed**
COLOR: **Metallic blue or green**
ANTENNAE: Threadlike
MOUTHPARTS: Chewing
EYES: Compound
WINGS: Hind wing with distinct lobe
LEGS: Walking
MISCELLANEOUS: **Coarse textured; when disturbed, roll up in a ball**
HABITAT:
Adults: Plants
Larvae: Same as host
FOOD:
Adults: Nectar
Larvae: Parasitoids or inquilines in nest of other hymenopterans

Cuckoo wasps are beautiful insects with a rough texture and a brilliant, metallic color. Their abdomen is hollowed out on the underside, allowing them to curl into a ball so only their tough outer skin is exposed to potential predators.

Cuckoo wasps are so named because they use the same technique for raising their offspring as the cuckoo bird: they lay their eggs in the nest of another species. Most are external parasitoids of full-grown wasp or bee larvae, but other species are cleptoparasites, which kill the host larvae so their own larvae can use the food stored in the nests.

Cuckoo wasp (note teeth at tip of abdomen)

Cuckoo wasp

Gall Wasps

Family: Cynipidae
(sy-NIH-pih-dee)

Characteristics:

SIZE: 1/16 to ¼ in. (1.6 to 6.4 mm)
SHAPE: Pronotum somewhat triangular in lateral view; **generally humpbacked**
COLOR: Dull brown or black
ANTENNAE: Threadlike; 11- to 16-segmented
MOUTHPARTS: Chewing
EYES: Compound
WINGS: Reduced venation
LEGS: Walking; first segment of the hind tarsi elongated
MISCELLANEOUS: Ovipositor arises anterior to apex of abdomen; **abdomen generally flattened sideways**
HABITAT:
Adults: **Host tree**
Larvae: **Galls on trees**
FOOD:
Adults: Often do not eat
Larvae: Gall formers; some inquilines; a few parasitoids of other insects

There are over six hundred species of gall-forming insects in the family Cynipidae. Most species produce galls on oak trees, roses, or members of the thistle family. These tiny insects are seldom seen by the casual observer and are difficult to identify, but the galls they produce are common and often identifiable.

Female cynipids lay their eggs in actively growing plant tissue, such as twigs, leaves, and even roots. The larvae emit chemicals that stimulate the plant to produce the gall, which serves as both shelter and food. The entire life cycle may take up to three years to complete.

Reproduction in cynipids is quite complex. In some species, fertilized eggs turn into females, and unfertilized eggs become males. In other species, no males are known to exist, and the eggs develop only into female wasps. Still other species have a complex type of alternation of generations. After mating, the eggs become females, which develop and lay eggs that can become either female or male.

Some cynipids are inquilines and have apparently lost the ability to produce galls. Instead, the female will lay her eggs in the gall of another species. The visiting wasp larvae feed on the available plant tissue, which may starve the original inhabitant.

all wasp

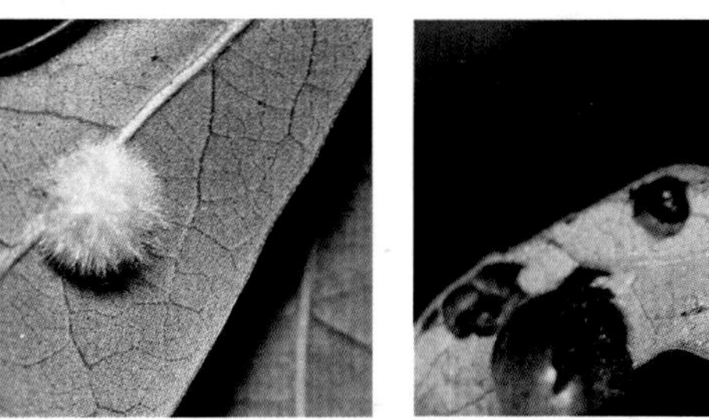

Gall on oak leaf

Galls on oak leaf

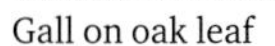

Ichneumon Wasps

Family: Ichneumonidae
(ihk-nyoo-MAH-nih-dee)

Characteristics:

SIZE: $\frac{1}{8}$ to $1\frac{1}{2}$ in. (3.2 to 38.1 mm)
SHAPE: **Abdomen sometimes laterally compressed; generally humped**
COLOR: Variable
ANTENNAE: Threadlike; **usually at least half length of body**; 16-segmented
MOUTHPARTS: Chewing
EYES: Compound
WINGS: Translucent
LEGS: Walking
MISCELLANEOUS: **Ovipositor rises in front of apex of abdomen, not capable of being withdrawn, often longer than body**

Ichneumonid wasp female (by Gary Brooks)

HABITAT:
Adults: Free living
Larvae: Host
FOOD:
Adults: Nectar
Larvae: Most internal parasitoids of immature insects

Ichneumonidae is a large family of wasps with nearly thirty thousand species worldwide and eight thousand in North America. They are similar to the braconids but are generally larger and have a different wing venation. Some females possess ovipositors that are longer than the body. Some parasitoids of wood-boring insects have ovipositors over three inches long. Only a few species can sting, but most will attempt to if handled.

Most species are internal parasitoids of Lepidoptera larvae and pupae and thus are considered beneficial. Some species, however, feed on the immature stages of beetles, and a few feed on spiders. Hyperparasitism is also common, in which ichneumons parasitize the larvae of other parasitoids, including other ichneumons.

Ichneumonid wasp male (by Gary Brooks)

Ichneumonid wasp

Mud Daubers, Digger Wasps, Thread-waisted Wasps

Family: Sphecidae
(SFEH-suh-dee)

Characteristics:

SIZE: $\frac{1}{4}$ to $1\frac{1}{4}$ in. (6.4 to 31.8 mm)
SHAPE: **Thread-waisted (subfamily Sphecinae); others quite variable**
COLOR: Black, blue, reddish brown, greenish; often metallic; some marked with yellow stripes
ANTENNAE: Threadlike
MOUTHPARTS: Chewing
EYES: Compound
WINGS: Present; not folded
LEGS: Walking
MISCELLANEOUS: Hairs on body few and not branched; **pronotum, short and collarlike; solitary**
HABITAT:
Adults: Soil; natural cavities
Larvae: Mud tubes; cells in soil
FOOD:
Adults: Nectar; ooze from ovipositor wound
Larvae: External parasitoids of insects or spiders

Sphecids are a large, diverse group of solitary wasps. The familiar thread-waisted wasps are easily recognized by their extremely slender first abdominal segment. Other species can be bright

Mud dauber wasp

Sphecid larva feeding on spider

metallic or dark with yellow stripes and are less easily identified.

All sphecids are parasitoids on a variety of other insects and spiders. Each species has a fairly specific host range. The cicada killer, *Sphecius speciosus*, is a large and conspicuous insect that will capture, sting, paralyze, and place cicadas in an earthen cell, where their larvae slowly consume the hapless victim.

Some sphecids make nests in the ground; others settle in hollow tubes, like water and air hoses. Still others construct mud tubes on the sides of various structures, such as the eaves of houses.

Some species can sting but are generally not aggressive.

Cicada killer, *Sphecius speciosus*

Sphecid wasp with caterpillar (by Gary Brooks)

Leaf-cutting Bees

Family: Megachilidae
(meh-guh-KIHL-ih-dee)

Characteristics:
SIZE: $\frac{3}{8}$ to $\frac{3}{4}$ in. (9.6 to 19.2 mm)
SHAPE: Bee shaped but stouter
COLOR: Generally dark colored; often with light bands on abdomen
ANTENNAE: Threadlike
MOUTHPARTS: Lapping; **tongue often elongated, flattened, and visible**
EYES: Compound
WINGS: Wings present
LEGS: Walking
MISCELLANEOUS: **Head larger in proportion to body than that of other bees**
HABITAT:
Adults: Flowers
Larvae: Nest in natural cavities or ground
FOOD:
Adults: Pollen; nectar
Larvae: Pollen; nectar; leaves

Leaf-cutting bee

Leaf-cutting bees are so named because the females of many species line their nest with pieces of leaves cut in neat circular patterns. They are quite similar in body shape to other bees but can generally be identified by the elongated and flattened tongue, dark color, and light bands on the abdomen.

Female leaf-cutting bees typically make nests in natural cavities such as tree holes, but some species prefer the soil. Pollen is collected and carried to the nest in a specialized area under the female's abdomen.

Most species are nonparasitic and solitary, but it is not uncommon to find the soil-inhabiting species in large communities.

Leaf-cutting bee

Leaf-cutting bee damage (by Gary Brooks)

Paper Wasps, Yellowjackets, Hornets

Family: Vespidae
(VEHS-pih-dee)

Characteristics:

SIZE: $\frac{1}{2}$ to 1 in. (12.8 to 25.4 mm)
SHAPE: Typically wasplike
COLOR: Most black, yellow, or red; often banded
ANTENNAE: Threadlike
MOUTHPARTS: Chewing; strong mandibles
EYES: Compound; inner margin notched by antennae
WINGS: **Wings folded lengthwise when at rest**
LEGS: Walking; hind tibia does not extend beyond the abdomen
MISCELLANEOUS: **Hind margin of pronotum strongly U-shaped**
HABITAT: Most social; a few solitary; **paper nest**

FOOD:
Adults: Predaceous on other insects, particularly caterpillars
Larvae: Predigested food from workers

Almost everyone who has ventured outdoors is painfully familiar with the paper wasp. These social insects build papery nests made from chewed wood, leaves, and other locally available materials. The numerous common species all share the characteristic of folding their wings lengthwise when at rest. Unfortunately, you sometimes must get dangerously close to observe that trait!

Polistes carolina is reddish and often referred to as "red wasp." *Polistes exclamans* is a common paper wasp in Texas but is often incorrectly referred to as "yellowjacket." Paper wasps build nests that are umbrella shaped and open underneath. Yellowjackets, such as the southern yellowjacket, *Vespula squamosa,* build an aboveground, papery nest that is virtually enclosed or in the ground.

All vespids are predators of other insects, particularly Lepidoptera caterpillars, and are considered beneficial—when in the right habitat!

Baldfaced hornets, *Dolichovespula maculate,* nest

Vespid wasp wings (note folded wings)

Paper wasp nest

Paper wasp with chewed caterpillar

Scoliid Wasps

Family: Scoliidae
(skoh-LEE-ih-dee)

Characteristics:

SIZE: $\frac{3}{4}$ to $1\frac{1}{4}$ in. (1.9 to 3.2 cm)
SHAPE: Typically wasplike
COLOR: **Usually black with yellow or reddish spots on abdomen**
ANTENNAE: Threadlike
MOUTHPARTS: Chewing
EYES: Compound
WINGS: **Many longitudinal wrinkles; dark**
LEGS: Walking
MISCELLANEOUS: Hairy
HABITAT:
Adults: **Flowers; plants**
Larvae: Host habitat
FOOD:
Adults: Nectar; pollen
Larvae: Ectoparasitoids of white grubs

Scoliid wasps are large, solitary wasps commonly observed on flowers. They have dark-colored bodies with yellow or orange spots on the abdomen and dark, wrinkled wings that do not fold lengthwise when at rest. Their bodies are covered with stout hairs.

Scoliids are sometimes referred to as "digger wasps" because the adult females burrow into the ground in search of scarab grubs (order Coleoptera, family Scarabaeidae). Once the grub is located, stung, and paralyzed, a cell is constructed around it, and the female lays an egg. A newly emerged larva attaches to the grub and begins to feed. Scoliids are considered beneficial because they often sting more grubs than necessary, and the paralysis is generally fatal.

Scoliid wasp

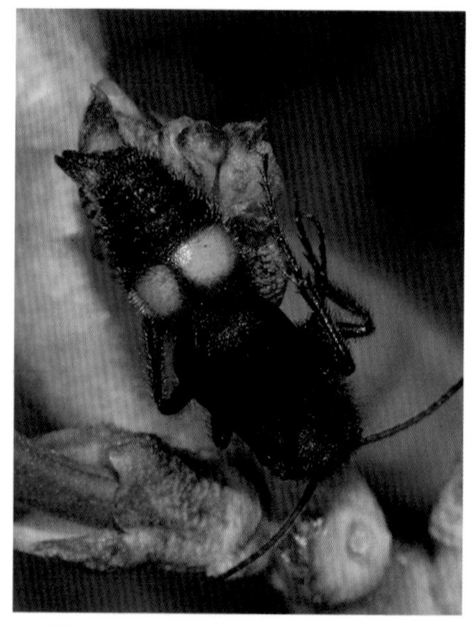

Scoliid wasp

Spider Wasps

Family: Pompilidae
(pahm-PIHL-ih-dee)

Characteristics:
SIZE: $\frac{1}{4}$ to 2 in. (6.4 to 50.8 mm)
SHAPE: **Typically wasplike**
COLOR: **Usually metallic black or blue; some with orange bands**
ANTENNAE: Threadlike; **female antennae often curled**
MOUTHPARTS: Chewing
EYES: Compound
WINGS: Not folded flat over abdomen; generally dark or smoky colored; **nervously twitch wings when searching for food**
LEGS: Walking; **slender; long and spiny; hind femora long, often extending beyond abdomen**
HABITAT:
Adults: **Open areas; on flowers or on ground**
Larvae: Burrows in soil constructed by adults
FOOD:
Adults: Nectar
Larvae: Parasitoids of spiders

Spider wasp

Spider wasps are commonly seen on the ground nervously twitching their wings and antennae as they search for prey. These insects also can be recognized by their elongated hind femora, which often extend beyond the tip of the abdomen, and by the females' curled antennae.

Spider wasps are solitary insects that generally construct an underground tunnel, often with several cells. When the female finds a suitable spider host, she will paralyze it and place it in one of the cells. She lays a single egg, typically on the spider's abdomen. The larvae feed on the paralyzed spider during their entire developmental period. Some species use the spider's own burrow to house the larvae and the paralyzed spider.

The tarantula hawk, *Pepsis* spp., is one of the largest wasps, yet it is still smaller than its prey. It is not clear how spider wasps can subdue their prey, but it appears that in some cases the spider may be somehow hypnotized. All spider wasps can inflict a painful sting.

Spider wasp

Stephanids

Family: Stephanidae
(stuh-FAN-ih-dee)

Characteristics:

SIZE: ¼ to ¾ in. (6.4 to 19.2 mm) (excluding tail)
SHAPE: **Cylindrical**
COLOR: Black
ANTENNAE: Threadlike
MOUTHPARTS: Chewing
EYES: Compound
WINGS: Present
LEGS: Walking; **hind leg femora swollen**
MISCELLANEOUS: **Head, round and with tooth-like crown; relatively long neck**

Stephanid wasp (by Gary Brooks)

HABITAT:
Adults: **Trunks of dead or dying trees**
Larvae: Host's habitat
FOOD:
Adults: Unknown
Larvae: Parasitoids of the larvae of wood-boring beetles and horntails

Stephanidae is a small group of insects that are rarely seen by the casual observer. These insects resemble ichneumons but are identified by their round heads, long neck, and the toothed crown. There are only six known U.S. species.

Typically, stephanids are found on dead or dying tree trunks, seeking the larvae of wood-boring insects. Their hosts include the larvae of wood-boring beetles and hymenopterans. The female stephanid locates the cell of the host, inserts her long ovipositor, and lays an egg. The developing larvae feed externally on the host and eventually pupate inside their host's cell.

Stephanid wasp (by Gary Brooks)

Sweat Bees

Family: Halictidae
(huh-LIHK-tih-dee)

Characteristics:

SIZE: $\frac{1}{10}$ to $\frac{1}{2}$ in. (2.5 to 12.8 mm)
SHAPE: Typically beelike
COLOR: **Black or metallic green or bluish**
ANTENNAE: Threadlike
MOUTHPARTS: Chewing
EYES: Compound
WINGS: Present; translucent
LEGS: Walking
MISCELLANEOUS: **Body not coarse textured; cannot roll into a ball**
HABITAT:
Adults: **Flowers**
Larvae: Soil
FOOD:
Adults: Pollen and nectar
Larvae: Mostly pollen; a few live off provisions of other bees

Sweat bee (by Jack Brady)

Sweat bee (by Jack Brady)

The halictids are a large group of common bees that are mostly black or brown. Some species, however, are bright metallic and resemble cuckoo wasps (order Hymenoptera, family Chrysididae) without the sculptured cuticle. The adults feed on pollen and nectar from a wide range of plant species and are often found in large numbers around blooming plants. For this reason, they are considered important pollinators.

Female halictids dig branching tunnels in shady, bare, flat ground or the sides of vertical banks. They provision each branch with pollen and lay a single egg. Although these are solitary insects, large numbers of females often nest in the same area.

Some species are attracted to perspiration, which explains the common name "sweat bees." These insects can be quite annoying and can sting when provoked.

Tiphiid Wasps

Family: Tiphiidae
(tih-FEE-ih-dee)

Characteristics:

SIZE: $\frac{1}{4}$ to 1 in. (6.4 to 25.4 mm)
SHAPE: **Elongated; slender**
COLOR: **Black with yellow stripes**
ANTENNAE: Threadlike
MOUTHPARTS: Chewing
EYES: Compound
WINGS: **Females wingless; males winged**
LEGS: **Females, forelegs modified for digging;** males, walking
MISCELLANEOUS: **Males possess upward-curved spine at tip of abdomen**
HABITAT:
Adults: Flowers
Larvae: Soil
FOOD:
Adults: Nectar
Larvae: Beetle larvae and mole crickets

Male tiphiids are elongated and have a rather large spine at the tip of the abdomen that curves upward, resembling a stinger. Although they appear quite fierce, they cannot sting and are harmless. Male tiphiids are common visitors to flowers, where they feed on pollen. Female tiphiids are wingless and burrow into the ground searching for the grubs of scarab beetles (order Coleoptera, family Scarabaeidae). She lays her egg next to the grub, and the larva feeds externally on its host. Tiphiids are solitary wasps.

Male tiphiid wasp

Velvet Ants, Cow Killer Wasps

Family: Mutillidae
(myoo-TIHL-ih-dee)

Characteristics:

SIZE: $\frac{1}{4}$ to 1 in. (6.4 to 25.4 mm)
SHAPE: **Resemble large ants**
COLOR: **Often brightly colored; reddish or orange; some with rings or spots**
ANTENNAE: Females, curved; males, straight
MOUTHPARTS: Chewing
EYES: Compound
WINGS: **Females wingless; males winged**
LEGS: Walking
MISCELLANEOUS: **Females very hairy and velvety looking; males less hairy; rapid runners**
HABITAT: Solitary; **prefer open, sandy areas**
FOOD:
Adults: Nectar
Larvae: Ground-dwelling bees and wasps

Velvet ants resemble very large, hairy ants, although they are actually wasps. They are usually solid red or orange but can have rings or spots on the abdomen; all-white species are also common. Female velvet ants are wingless, but the males are winged and less hairy.

Velvet ants are solitary insects that prefer open, sandy areas associated with their prey. These insects are active during the daytime; they are rapid runners and seldom stop to rest. Female velvet ants seek soil inhabited by larvae of other wasps and bees to lay their eggs. The developing larvae are ectoparasitoids, feeding on the host larvae, and eventually pupate in the host's underground cell. There is usually one generation per year.

As the name "cow killer" suggests, the female velvet ant can inflict a very painful sting; however, there are no actual reported cases of livestock being killed as a result of their sting. The males have no sting and are harmless.

Female velvet ant

Male velvet ant

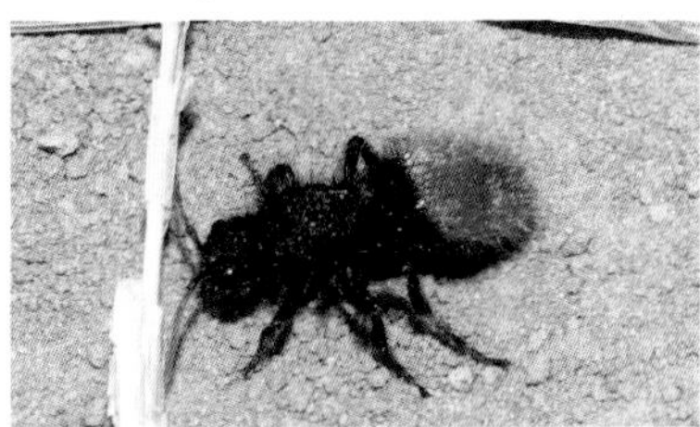
Female velvet ant (by Gary Brooks)

Glossary

ABDOMEN The hindmost portion of an insect's body.

AMETAMORPHOUS A form of metamorphosis in which the immature resembles the adult stage but smaller; stages of development include the egg, nymph, and adult.

APOSEMATIC COLORATION Colors or patterns used by animals to indicate the presence of toxic chemicals.

BIOCONTROL The use of naturally occurring enemies to control a pest.

CEPHALOTHORAX The combined head and thorax of certain arthropods.

CERCUS (pl., cerci) Tail-like appendage on some insects.

CHELICERA (pl., chelicerae) Pincherlike mouthpart of some arthropods.

CHITIN The tough, insoluble chemical found in the exoskeleton of arthropods.

CHRYSALIS (pl., chrysalids) The aboveground pupal case of butterflies.

CLEPTOPARASITE A parasite that feeds on the food reserves of another insect.

COMPLETE METAMORPHOSIS A form of metamorphosis that includes the egg, larva, pupa, and adult stages.

COMPOUND EYE The insect eye composed of individual hexagonal facets.

CORNICLE Structure on the abdomen of aphids that secretes defensive chemicals.

COXA (pl., coxae) The basal portion of an insect leg.

CREMASTER The hooklike structure at the end of a butterfly pupa.

CROCHET The hooklike structure at the end of the proleg of lepidopteran caterpillars.

CUNEUS (pl., cunea) A triangular-shaped structure on the forewing of some heteropteran insects.

DORSAL Located at the top.

ECTOPARASITE A parasite that feeds externally.

ECTOPARASITOID A parasitoid that feeds externally.

ELYTRA The forewing of coleopterans.

Endoparasitoid A parasitoid that feeds internally.

ENTOMOLOGY The study of insects and related arthropods.

EXOSKELETON The external bone structure of arthropods.

FACET A lenslike division of the compound eye of insects.

FAMILY The level of the taxonomic hierarchy between the order and genus.

FEMUR (pl., femora) The segment of the leg located above the tibia.

FOREWING One of the pair of wings located on the mesothorax.

FURCULA The forked tail of springtails.

GALL Abnormal plant growth resulting from insect infestation.

GENITALIA External components of the reproductive system of insects.

GENUS (pl., genera) The level of the taxonomic hierarchy between the family and species.

GRADUAL METAMORPHOSIS A metamorphosis type that includes the egg, nymph, and adult stages.

HALTERE The knoblike metathoracic wing of dipterans.

HAMULUS (pl., hamuli) A hooklike structure on the hind wing of hymenopterans that attaches to the rear margin of the forewings.

HIERARCHICAL NOMENCLATURE SYSTEM The system of classifying organisms from broadly similar to very specifically related.

HIND WING One of the second pair of wings located on the metathorax.

HYPERPARASITE An insect that parasitizes a parasite of another organism.

INCOMPLETE METAMORPHOSIS A type of metamorphosis that includes the egg, an aquatic naiad, and adult stages.

INQUILINE Pertaining to an animal that lives in the nest of another animal.

INSTAR A stage of an insect's development between molts.

LARVA (pl., larvae) The active, immature stage of insects with complete metamorphosis.

MESOTHORAX The second or middle thoracic segment.

METAMORPHOSIS The development of an organism from egg to adulthood.

METATHORAX The third and most rearward segment of the thorax.

MOLT A phase of development in which the insect sheds its exoskeleton.

MORPHOLOGY The study of the physical structures of an organism.

MYIASIS Infestation of fly larvae.

NAIAD The aquatic, immature stage of odonatans and ephemeropterans.

NOCTURNAL Active at night.

NYMPH An immature stage that resembles the adult stage but without fully developed wings.

OCELLUS (pl., ocelli) A simple eye common on many insects.

OMNIVOROUS Pertaining to an organism that will eat a wide range of foods.

OOTHECA (pl., oothecae) The egg case of mantids and cockroaches.

ORDER The level of the taxonomic hierarchy between the family and class.

OSMETERIUM (pl., osmeteria) A gland on the head of certain lepidopteran larvae that can be everted.

OVERWINTER To survive the winter through a period of dormancy.

OVIPOSITOR The egg-laying apparatus.

PALP A fingerlike projection that is part of the mouthparts.

PARASITE An organism that feeds on other organisms, causing harm but typically not killing the host.

PARTHENOGENESIS A form of reproduction in which the embryo develops without being fertilized.

PEDIPALP One of the second pair of appendages on the cephalothorax, which is often modified to form pinchers.

PHYLUM The level of the taxonomic hierarchy between the kingdom and class.

POLYORNATISM The occurrence of several distinct color patterns within a population.

PREDACEOUS Pertaining to an organism that kills and eats another organism.

PRIMARY PARASITOID An organism that, during the immature stage, feeds on another organism and kills the host upon emerging as an adult.

PROBOSCIS The beak or extended mouthpart of some insects.

PRONOTUM The first thoracic segment.

PROTHORAX The front section of the thorax.

PUPA (pl., pupae) The nonfeeding stage of development of insects with complete metamorphosis.

PUPATE To change from the larval to the pupa stage; occurs in insects with complete metamorphosis.

SCAVENGER Organism that feeds on decomposing animal and plant matter.

SCUTELLUM A triangular structure at the base of the wings of some insects.

SECONDARY PARASITOID A parasitoid of a primary parasitoid.

SOLITARY Pertaining to an organism that lives alone.

SPECIES A group of organisms that can mate and produce viable offspring capable of reproduction.

SPIRACLE An external opening that insects use to breathe.

SUPERFAMILY The level of the taxonomic hierarchy between the order and family.

SYSTEMATICS The study of the diversity and relationships of life.

TARSAL FORMULA The number of tarsal segments on each leg; typically distinctive for each insect family.

TARSUS (pl., tarsi) The last portion of the leg; analogous to a foot.

TAXONOMY The science of classifying organisms.

TERRESTRIAL Pertaining to an organism that lives on land.

TERTIARY PARASITOID A parasitoid of a secondary parasitoid.

THORACIC LEG Leg located on the thoracic segments of a larva.

THORAX The middle body segment of an insect that bears the legs and wings.

TIBIA (pl., tibiae) The section of the leg between the femur and tarsus.

URTICATING SPINE The spine of some lepidopteran larvae that is hollow and contains irritant chemicals.

VENTRAL Pertaining to the lower or underside of an organism.

Index